Follow the Chemistry:
Lure, Lore and Life

Follow the Chemistry:
Lure, Lore and Life

An Autobiography of
Goh Lai Yoong

World Scientific

NEW JERSEY · LONDON · SINGAPORE · BEIJING · SHANGHAI · HONG KONG · TAIPEI · CHENNAI

Published by

World Scientific Publishing Co. Pte. Ltd.
5 Toh Tuck Link, Singapore 596224
USA office: 27 Warren Street, Suite 401-402, Hackensack, NJ 07601
UK office: 57 Shelton Street, Covent Garden, London WC2H 9HE

British Library Cataloguing-in-Publication Data
A catalogue record for this book is available from the British Library.

FOLLOW THE CHEMISTRY: LURE, LORE AND LIFE
An Autobiography of Goh Lai Yoong

Copyright © 2010 by World Scientific Publishing Co. Pte. Ltd.

All rights reserved. This book, or parts thereof, may not be reproduced in any form or by any means, electronic or mechanical, including photocopying, recording or any information storage and retrieval system now known or to be invented, without written permission from the Publisher.

For photocopying of material in this volume, please pay a copying fee through the Copyright Clearance Center, Inc., 222 Rosewood Drive, Danvers, MA 01923, USA. In this case permission to photocopy is not required from the publisher.

ISBN-13 978-981-4304-00-9
ISBN-10 981-4304-00-X

Typeset by Stallion Press
Email: enquiries@stallionpress.com

Printed in Singapore.

Dedication

*To my Parents, Husband, Children,
and all those who have touched my life.*

Contents

Prologue		ix
Acknowledgements		xiii
Chapter 1.	Early Memories	1
Chapter 2.	Start of School	9
Chapter 3.	On The Move — An Unsettling Two Years	13
Chapter 4.	Back To Normalcy	21
Chapter 5.	Pre-University and University	29
Chapter 6.	In the University of Malaya, Kuala Lumpur	41
Chapter 7.	Postgraduate Studies in London	47
Chapter 8.	Beyond the Ph.D. and Chicago	63
Chapter 9.	A Beginning in Academia	71
Chapter 10.	Sabbatical Leaves	79

Chapter 11.	Research Reality in the Home Environment	91
Chapter 12.	Overseas Conference Participation and a Bit of Travel	101
Chapter 13.	In Service of the Profession	115
Chapter 14.	My First Retirement and a Doctor of Science	119
Chapter 15.	A Second Retirement and Beyond	125
Chapter 16.	Family	131
Epilogue — Reflections		137
Appendix I	List of Acronyms	145
Appendix II	Selected List of Journal Publications	147
Appendix III	Published Conference Abstracts and Proceedings	157

Prologue

It was one morning in April 2009 that Leung Pak Hing conveyed to me the message that Lee Soo Ying and he would like me to write up my story. My initial response was 'surely that's a joke! I do not have much to tell'. Only sometimes, perhaps in rare moments of false inspiration, did I mentally moot the idea of writing up about my family — 'Three generations in Malaysia' — when I retire. But, an autobiography? That had never crossed my mind, not even in my dreams. First and foremost, I have always been and still am only a low-profile university teacher. Next, I am a very private person, describable as an ordinary person living a most ordinary life. Only famous personalities, e.g., world and national leaders, politicians, CEOs and eminent people of their professions, record their important experiences for posterity. Thus, we have Hilary Clinton's *A Living History*, Barack Obama's *Dreams from my Father*, Benazir Bhutto's *Daughter of the East*. Less dynamic characters than these would take one to two years off in their retirement to deliberate, before putting down their stories in print. In Bhutto's case, posterity is lucky that she wrote up her extraordinary tale so early, because as it turned out, she would never have a retirement period in which to do it. Eminent others whom I know quite well have written up their memoirs in retirement or semi-retirement, e.g., Rayson L. Huang in *A Life Time in Academia* and Ho Peng Yoke in *Reminiscence of a Roving Scholar: Science, Humanities and Joseph Needham*.

To consider to write up my story in a few months, in the midst of teaching commitments, without the tranquility of mind to take a stroll down memory lane and to deliberate, I was not sure if I could produce a good coherent tale. But maybe it will be do-able, as I did not think I have much to tell anyway! But I need to know the purpose, and I will need to reconsider, also consult with my husband, as inevitably part of his story will be revealed as well. Soo Ying sent me samples of biographies of some women Nobel laureates! But really I am far, far from such calibre even if I was allowed to develop in a right environment during the earlier days.

After noting that my story is intended for the motivation of local women into science if it can be of any help, and after some persuasion, I agreed to write, but only a simple tale as my husband suggested and later helped to edit. The reader will note that my career path was a drift along a Providence plan. My only efforts seemed to make sure I kept afloat by doing what was expected of that providential career. I hope that this account would encourage more women into chemistry or at least into the sciences, as the concern remains strong regarding the low representation of women among colleagues in Science, Engineering and Technology, despite a decade-old attempt at improving the situation. Readers may wish to view updates at the site: www.thefreelibrary.com (article posted 8/17/2009: *Is there a global warming toward women in academia?*) Equally important, this account is meant to be a tribute to the

Prologue

many special people — relatives, friends and several members of the Chemistry community — who have helped me in some way or have created an impression throughout the journey of my career life.

> *To pursue the perfect in life*
> *Thoughts spring never to die*
> *Unending hope forever alive*
> *If only with dreams to live by*
> *Surely only time and life flies*
> *Other things may be but lies*
>
> (captured by S. H. Goh)

Acknowledgements

This autobiography is a tribute to all the people who have crossed my path, ranging from family and extended families to all my teachers and lecturers, the mentors of my graduate and postgraduate studies, the American and Australian professors who had hosted me for sabbatical and research leaves, and colleagues and students at three universities (UM, NUS and NTU).

I thank my husband Swee Hock for his ever-consistent support and his careful critical editing and suggestions for improvement of this write-up, and reading through the entire manuscript. I must congratulate my children for developing self-reliance in play as well as in their studies, thus allowing me to adhere to the demands of the profession.

Writing this account while in the service of the Division of Chemistry and Biological Chemistry (CBC), School of Physical and Mathematical Sciences (SPMS) in the Nanyang Technological University of Singapore, I wish to express a special note of thanks to Lee Soo Ying (LSY), Leung Pak Hing (LPH) and Loh Teck Ping for their support here in my third retirement position, and again to LSY and

LPH for their suggestion, persistent encouragement and support in bringing this book to fruition.

In addition, I would also like to thank Gladys Lee Chai Fen, Chai Siew Ping and Xu Chang (from CBC) for organizing the pictorial pages, and World Scientific Publishing Co., especially Lim Sook Cheng, V. K. Sanjeed, Loo Chuan Ming and Xiao Ling.

CHAPTER 1

Early Memories

It was in the small town of Ipoh, in a rich tin-bearing valley of the Kinta River in Perak of British Malaya, where my first recollection of anything began with a rude awakening in the dead of night by anxious adult chatter, my maternal grandmother's excited voice being distinct. The drone of invading Japanese airplanes was obvious as they flew low, and the adults were discussing about taking shelter in the tunnel below. Unfortunately water had collected in it, so no one went in. Although too young to understand the implications of all this, I could sense an element of fear, but fortunately there was no tragedy in that countryside. It was years later that I became aware that our whole extended family was then taking refuge from Japanese invaders in the home of a wealthy tin-miner friend of my parents and grandparents in the village of Tronoh (see map overleaf). Strangely, I have absolutely no recollection of the house or the environs. On that same occasion, someone mentioned that tiny maggots were growing in the rice stored for a long period of time in view of scarcity. For a young impressionable mind, this incident had created a lasting paranoid fear of creeping and crawling creatures even though those are just common weevils.

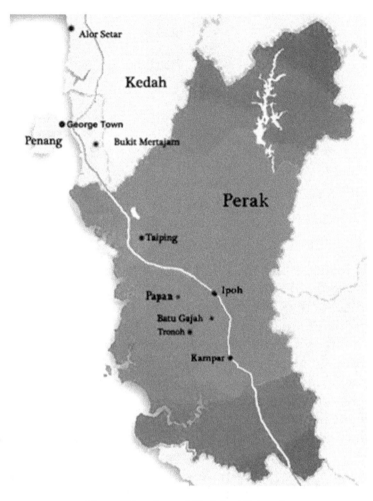

Map of Perak and Kedah, Malaysia

The name Ipoh derives from a local Ipoh tree, *Antiaris toxicaria*, the sap of which is poisonous and was used by Orang Asli (indigenous people) to coat the tips of the darts of their blowpipes. The town, the capital of the silver state Perak (silver in Malay), has variously been called 'The Town That Tin Built' or 'City of Millionaires', referring to the vast fortunes made during the boom of the tin-mining

Early Memories

industries. Sadly its growth had stagnated with the demise of the tin industry in the late 1970s, resulting in the migration of many young talents to cities like Kuala Lumpur and Singapore, so much so that Ipoh has since been referred to colloquially as a 'sleepy hollow', a good place for retirement, even though the city and its suburbs are expanding all the time. Tronoh located some 30 km south of Ipoh was also once a thriving tin-mining town famed for its deep-shaft mine. Today Tronoh is also a sleepy little town surrounded by oil palms although that may change, now that two universities — Universiti Teknologi Petronas (founded in 1997) and Universiti Teknologi MARA (founded in 1996) — are in the vicinity. Nearby Ipoh is the town Kampar, which is now home to Universiti Tunku Abdul Rahman or UTAR, originally set up as TAR college, offering external degrees, e.g. London University External and Campbell University of U.S.A., for students who failed to obtain admission into the national public universities.

Here is a little 'aside' to my ancestral forebears. I was told that my grandparents were first generation migrants to then Malaya from Guangdong Province in China; my paternal grandfather was a Chinese physician, with a hand in tin-mining as well, while my maternal grandfather was partly a businessman of sorts and partly a tin-miner. They plied to-and-fro between then Malaya and China, returning to China when times were trying in Malaya, such as during the Great Depression of 1929. My mother recalled that the British government of the time encouraged Chinese residents to return to China with paid sea fares. This obviously was very disruptive for my mother's education, but anyway that was not considered essential for girls in those days. My mother attended private classes under the masters of Chinese language and literature in Ipoh; one of these masters was Ho Tih Ann, the father of Professor Ho Peng Yoke (physicist-*cum*-renowned historian of Chinese Science), who had held professorial positions in UM, Griffith

University and HKU, and Directorship of The Needham Research Institute of Cambridge University (see his autobiography, *Reminiscence of a Roving Scholar: Science, Humanities and Joseph Needham*, 2005, World Scientific Publishing.) My mother's cousin, aunt Yuet was not given any formal education, but incredibly managed to self-educate herself sufficiently in Chinese to be able to read novels and newspapers. My two sets of grandparents were jolly good friends and had hand picked each other's children for daughter- and son-in-law, respectively.

My father (Chung Yew) and three of his six siblings (two brothers – Chung Weng and Chung Tai and a sister Chuk Ling) managed to survive all manner of tropical diseases without antibiotics. Together with his elder brother Chung Weng, he finished well in Anderson School (one of Ipoh's distinguished schools, now just past its centenary) and furthermore he secured a Queen's scholarship to study medicine in UK. However, as fate would have it, this was recalled because of the uncertain climate during the throes of the Great Depression and an impending war, already looming in Europe. Hence his next best option was to go for free university education in Sun-Yat-Sen University in Quangzhou, China, from where he obtained a Bachelor's degree in Civil Engineering. I had made it a point to visit that campus, when I accompanied my husband to an '*Oils-and-Fats Conference*' in Guangzhou in 2007. During the final year of his university course he married my mother, the marriage being brought forward on account of my grandfather's imminent death (at age 49) from kidney disease arising from kidney stones. After graduation, they tried to settle down in China, my father starting work in Yunnan in 1937. It was a harsh environment there and my mother could not tolerate the living conditions there and fell very ill. So they returned to Malaya, where my father secured a government job in Penang with the Public Works Department (PWD). Their next move was a transfer to Ipoh, hometown of my maternal grandparents. Unfortunately for my father and indirectly for the family, degrees from Chinese universities were not recognized by the British government, and so my father had to settle for a position of technical assistant, below his qualification status, subordinate to a British principal engineer. It was in Ipoh where my parents settled and brought up their family of six children, amongst whom I am the eldest.

Other than that vivid recollection of the night in Tronoh of low-flying Japanese warplanes, memories of my earlier years are scant in detail. I only remember skeletons of episodes in that little semi-detached government bungalow in Greentown, Ipoh. My teenage aunt (Yan Sin) had to dress up as a boy on occasions to be 'safe' in those days of Japanese occupation; in my childhood naivety I thought that was to avoid being beaten up or attacked because boys were supposedly stronger. It was later that I was not spared some recounts of atrocities committed by the Japanese soldiers. Years later in secondary school, we learnt of the Ipoh heroine Sybil Kathigasu, nurse and wife of a doctor, arrested and horrendously tortured by the Japanese Military Police, the Kempeitai, for giving medical treatment to the anti-Japanese resistance fighters hiding in the hills of Papan, a small town near Ipoh. She became the only Malayan woman ever to receive the George Medal, a British civilian award for bravery, before she died in 1948 (at age 49) from injuries sustained (broken skull, jaw and spine and paralysed legs), despite numerous operations in Britain to fix her broken body. Her story is engraved in the heritage of Perak; her memoirs finished just before her death, the ending sections by dictation when she could not write anymore, was published posthumously in 1954, and recently re-published to mark the sixtieth anniversary of her death. (*No Dram of Mercy*, Sybil Kathigasu. ISBN 983-2197-22-8, Prometheus Enterprise, 2006).

Indeed, those were times of hardships and scarcities. Many like the Kathigasu's had to farm for food. I remember we also grew some of our own food in the limited land around the house. My father's cousin (on his maternal side) who was living with us with his newly-wed wife, once harvested buckwheat from our small front yard and we were overjoyed, awaiting the biscuits to be made therefrom! One day my mother gave me a hard-boiled chicken egg, a relatively rare food item, and I was to learn that it was because of my birthday. Come to think of it, I cannot remember having chicken meat for regular meals during my early school years. Pork was relatively more available, but chicken meat was a luxury reserved only for festive days. This was mainly because chickens were difficult to breed in those days when vaccines were unavailable. Because of that, the few chickens in our yard were closely watched, and slaughtered as soon as they refused to feed — a prelude to sickness — to which they would all succumb one after the other in a matter of days, much like, though not as serious as, the present day H5N1 contagions. In those days of food scarcity, it was no wonder that our maid coveted a treasured chicken egg, but she was spotted hiding it in an unused charcoal stove by the wife of my father's cousin. Not knowing about kleptomania in those days my mother had to ask her to leave for that and also upon finding that she had hidden bags of buckwheat among her belongings. In one of those days, I had flung away a sweet potato as far as I could, as a worm crawled out while I was scraping it. I have a vivid recall of my father coming home to relay the news that the Japanese had surrendered. We learnt that there was much rejoicing in town because the war had ended. The next few days, my father told of bags of Japanese 'banana' currency discarded in the streets.

We were very fortunate to have survived the war years but other trials remained. My brother Cheong, two years my junior, contracted malaria which was to weaken him

Early Memories

so badly that any slight attack of fever would bring on an onslaught of fits. I used to watch in fear, as my mother tried to force a spoon between his teeth to prevent him from biting his tongue. Indeed, my mother kept tender loving watch on him every moment of his feverish days. Such incidents persisted till his school age. Having been accustomed to such protective care, his first days in school must have been clouded with fear, and like some other boys, he ran homewards once an accompanying aunt was out of sight during the first days, bringing his teacher running after him.

CHAPTER 2

Start of School

After the Japanese had left, life began to return to normal. One evening, I overheard my parents discussing school options for me, and deciding that the Convent of the Holy Infant Jesus (CHIJ) was the best for girls, as it was an English school noted for strict discipline of girls, a school which trained and expected its girls to follow its motto '*Simple dans Ma Virtu, Forte dans Mon Devoir*': Simple in Virtue, Steadfast in Duty. So I was registered there.

My first days at school were nerve-wrecking, as I was quite 'unprepared'. I felt like I was being thrown among the wolves, as I was untutored in English. The only pre-school tutoring I had was when my father tried to teach me Confucian Chinese, but such classical texts were beyond the comprehension of my then obviously immature mind, and my father eventually dismissed me as 'such a stupid girl' and had forsaken efforts to educate me further in the Chinese classics. In the classroom, everyone else seemed conversant with the English language and could sing 'foreign songs', of which I was absolutely ignorant. On one of those days, as my classmates were happily singing along, I was overcome by confusion and anxiety, and I suddenly fainted. The resulting commotion must have made the teacher really worried! Several days later, my caring teacher passed my house as a pillion rider on a motor-bicycle but I ran indoors to avoid her, not wanting her to inform my mother of my shocking incident, of which my parents never came to know about. Coincidentally, my sister Lai Kien told me that she also had a similar incident to tell!

Not long after, the 'clever' girls (though I learnt later, they were actually over-aged girls, whose schooling had been delayed or interrupted by the War) were removed to another class, and I had the chance to settle down in a normal learning environment, in which I could finish up as eleventh in the class in that half year of Primary 2. At that time, school years went from Primary 1 and 2, through Standards 1–6, to Forms 1–5 with optional pre-university Form 6 (Lower and Upper) based on some selection criteria. In Standard 1, school was more enjoyable and I made up my mind to enjoy it to the fullest by neglecting all homework, whatever the consequences. Most surprisingly, at the end of term, when D-day arrived for the Principal, the Reverend Mother, to come to give us our results, she called out my name and I thought it was to reprimand me for having performed poorly. Actually, I could barely understand that she was in fact congratulating me, in her heavily French-accented English, for being first in the class. What a surprise indeed, much like a dream. This had a profound effect and the dream had just begun; so I was first in class without any effort — here I must record that my father had helped me with vocabulary, synonyms and antonyms, etc. — so was *first* where I should strive to remain? At the end of Term 2, I was second, gaining first again at the end of the year. Then came unfortunate events and I had to leave school; that was at the end of 1947, to be followed by almost two years of upheaval in the lives of all in our family.

But before that, just a little interlude about activities in after-school hours and during school vacations in those two to three years. There were little 'jobs' in the house

Start of School

which I was taught to do. One was to convert fresh milk, bought from a neighbouring Sikh vendor, to 'stable' sweetened condensed milk for no-refrigeration keeping, for my baby brother Kai Fatt. The continuous stirring over a low flame was quite fun for me, especially, as I watched the final thickening process — an early introduction to chemistry! Likewise, I also took delight in making baby food — a gluey paste from rice flour and a few other ingredients. My other (weekly) assignment was the cleaning of the family ancestral altar. At other times, it was children's play, which was mostly homely and unsophisticated; our games consisted of 'figure of eight jumping', rope-skipping and simple games using pebbles, with my brother Kai Cheong and two neighbourhood children. At school, we were taught the various types of stitching for a hem or a seam, and simple cross-stitching to sew alphabets on a handkerchief, or designs on a tablecloth. The kind teacher, a young girl, would gather together in her home the students in her neighbourhood to teach us more, especially during the school vacations. During these inter-term breaks, my mother also prescribed a lot to do. She taught me embroidery of the tapestry type, already begun before my schooling days; she trained me to perfection — by having me unpick whenever the 'underside' was messy! Eventually, I could produce an embroidered pillow case in a few days, though the head-panels on door curtains took a while longer. I would hurriedly finish off a piece, in order to get on to play; but then I was given another piece to do, for instance, when a couple of our family friends came visiting, saw and requested 'please sew one for me, it's so beautiful!' Thus went my whole vacation. But who is to say that this sort of girls' training had been to inculcate patience, perseverance and an eye for fine detail useful for Chemistry in my later life! I had wondered about this and did teach my children some embroidery in their childhood years. In dress-making my mother was quite remarkable as without any training, she could examine a dress or pair of trousers and be able to make the cutting to sew a replica, and that's how she sewed all our clothes, but always making them oversize so that they fitted us for a longer time. Those pictures of brother Cheong in his odd baggy trousers are still around. Sewing was by means of a foot-paddled sewing machine and I used to stand behind my mother and observe. I was already introduced to using the sewing machine barely two years into schooling; again my mother insisted on perfection — this time by having me unpick crooked seams — and I suppose that's how it was ingrained into me very early that everything had to be done well. As an aside, I was actually very surprised when some of my students in NUS told me that they have never used a sewing machine, because there was none in their home. No wonder then, there was a lucrative sewing service stall in the West Coast market area (close to Kent Vale, NUS staff apartments), catering to simple jobs like mending, sewing of hems and buttons!

CHAPTER 3

On The Move — An Unsettling Two Years

Coming back to the main course of historical events, my father was transferred to a very small town, Alor Star, the capital of the northern state of Kedah, on a job promotion. My mother had to manage the moving — lock, stock and barrel — in addition to caring for six children. Tending to the daily needs of a brood of six young ones must have been very stressful, with my youngest brother Kai Fatt barely one year old. Fortunately, she could enlist the assistance of aunt Yuet, who was unmarried and was housekeeping for a family in Singapore, to accompany us. In the hustle and bustle of things, my mother broke a mirror while taking it down from a wall. I remember being very fearful as I watched, having read in fairy tales of seven years of bad luck following such events, and coincidence or otherwise, we did suffer some of that in the next few years! The day for good-byes came, and we started the day's long steam-engined train journey to Alor Star, accompanied also by a cousin brother of my mother (Uncle Leong). The journey carried us through continuous flat expanses of green ricefields, prevalent in the whole state of Kedah — a monotonous scenery till we reached our destination. In Alor Star, the PWD was short of quarters for its employees, and unable to arrange for any sort of decent rentable private housing in town. In an adjoining kampong (meaning: village), my father found and rented a semi-detached wooden house with a detached 'backyard toilet'. Such inconveniences were common in rural environments in those days. I recall feeling very sad when Uncle Leong left us after a week to return to the relative comforts of Ipoh.

Then came schooling in this new place for me and brother Kai Cheong. We were admitted into government schools. I felt very uncomfortable in mine, in an environment very different from that in Ipoh's CHIJ, run by the religious sisters. As it happened, I could read (English) much better than the girls in my new school, and the unnecessary praise and comparison lavished by the teacher made me more uncomfortable. I remained wary of praise but would have been happy with just a light encouragement, which most of the time I did not need. Being really miserable in that school, I used to stand in a corner to do my own things at recess time. Even at that tender age when I could feel concern about my father's health, I sensed that brother Cheong must have felt the same, though we never compared 'notes'. It had only been a couple of months when my father, who had stones in one of his kidneys, found that infection had set in. However, high fever and body shivers caused the doctor to misdiagnose as an onslaught of malaria, prevalent in those days. But my parents knew otherwise as my father appeared to be physically incapacitated with wrong treatments. Unable to obtain proper medical care in a relatively rural environment, as well as appropriate medical leave, my parents made the decision that he should resign and return to Ipoh for treatment. So it was that we

On The Move — An Unsettling Two Years

moved — lock, stock and barrel — again, back to Ipoh after only three months in Alor Star.

In Ipoh my father had accommodation in a hotel, with my mother attending to him, while the rest of us 'squatted' in one room in the second storey of a shophouse, a bookshop at 39 Theatre Street, a pre-war shophouse leased to my grandparents by a wealthy landlord residing in the town of Kampar. My father was treated by one Dr. Ng Yoke Hing, a Queen's scholar who had just returned from UK as an MRCP specialist. On a couple of nights he had conscientiously attended to my father till the wee hours of the morning. His dedication temporarily saved my father's life, but surgery was still necessary. On his recommendation and after arrangement for surgery in Hong Kong was made by my paternal uncles, Uncles Chung Weng and Chung Tai, our whole family was on the move again to Hong Kong. My mother was uncertain about my father's chances of survival and in case of any eventuality she wanted us to rely on his extended family in Hong Kong, because in Malaya, she only had a married sister and two cousin brothers. If things had turned out badly some of my younger siblings would have been given away to relatives to be raised and this was a frightening thought, hopefully to remain only as a bad dream.

So in August of 1948, we together with Aunt Yuet accompanying us, went by train to Singapore, there to embark on a ship to Hong Kong. My invalid father was checked into a Class 2 cabin, while the rest of us had to make do in a simple Class 3 berth — an expansive borderless space with a continuous raised wooden platform, skirting the hull of the vessel, leaving a large expanse of empty space in the centre. The passengers all slept on the platform, with luggage kept by their side. Being in the hull, it was windowless and the atmosphere was hot, humid and stuffy. The lack of ventilation aggravated the misery of sea sickness; my siblings and I could not partake of food, which we had to collect from another section of the ship. As a result we all felt sickly and weak. I cannot recall about washing and toilet facilities, which must have been minimal. Every now and then, in fact, whenever permitted, we would go 'upstairs' to our father's cabin level where the fresh sea-breeze helped us get some reprieve from sea-sickness. So four or five days passed, but it seemed so long, and I remember next being helped into a swaying sampan on arrival to get on land in Hong Kong.

In Hong Kong, my father's siblings lived in Shamsuipo, (Uncle Chung Weng), and Hong Kong island (Uncle Chung Tai and Aunt Chuk Ling). For convenience, we went first to Uncle Tai's apartment, a small apartment adjoining one of the spacious large castle-like mansions in uptown Hong Kong. After discussion between my father and his siblings, it was decided that while my father underwent surgery, brother Cheong and I were to live with Uncle Tai, who would prepare us for school admission by tutoring us in Chinese (Cantonese), the medium of instruction in most primary schools. Uncle Tai is a post-war B.Sc. Honours graduate of the University of Hong Kong and was a mathematics teacher in Wah Yan College. The rest of my four siblings with Aunt Yuet were to go to live separately in my maternal grandparents' (Yip family) home in Shunde of Mainland China. My grandmother had returned to live there in 1946 after WWII since my grandfather had passed away during the war for lack of appropriate medical treatment. So after two days in Hong Kong, they left us, my mother accompanying but returning soon to Hong Kong to attend to my sick father. Only lately did my sister Lai Kien relate to me about hearing gunshots fired by bandits or soldiers at the ferry during the crossing to mainland China. Aunt Yuet was to remember the ordeal for the rest of her life, especially piggy-backing little brother Fatt for the whole journey. Lai Kien was old enough to remember very

On The Move — An Unsettling Two Years

well her rich experience of life in the Yip family home, especially the social interactions with our second cousins living there.

Soon it was time for my father's surgery in St. Paul's Hospital in Causeway Bay. Two competent European surgeons performed the operation, which took several hours, during which time Uncle Chung Tai was in prayer in the hospital chapel, as my mother later related to us. My father's infected kidney could not be saved and had to be removed, so he survived on the other one for the next thirty-five years. A second operation was found necessary a couple of months later. After his condition had stabilized, my mother left again for the Yip family home in Shunde. In Hong Kong, Uncle Chung Tai coached Cheong and myself in Chinese and even pampered us, while my aunt and grandmother had to tolerate the doubling of the occupancy in their little apartment. It was the first time that we were treated to movies in an a/c theatre, as air-conditioning had not arrived in Malaya; it was also the first time that brother Cheong was taken to watch basketball matches. Good news was when we passed the entrance examinations of two premier schools in Hong Kong.

China was getting close to civil war as Mao Zedong's army took control of much of the country. However, my recuperating father had not been keeping up with the news. Television had not arrived. It was difficult for us to believe the dream/vision he related one morning — that his late father appeared to him and asked him to send for my mother from China, or she would not be able to leave because of an imminent Communist takeover. So he did by letter, as there were no telephone or telegram services in those days. Apparently in the remote village, there was little news of current events and everything appeared blissful. With a small packed bag and carrying toddler brother

Kai Fatt my mother had to catch a ferry from Canton (present day Guangzhou) with a cousin brother accompanying her. As my mother recounted, conditions in Canton were totally chaotic — there were no more boats for the crossing; if any they were all full. Finally her cousin found one 'vacancy' but without seating after paying some 'coffee money' (a tip, socially acceptable camouflage for bribery), it was the last ferry to cross to Hong Kong. On arrival, my mother related the risky trip — sitting on her luggage and carrying Kai Fatt, she and fellow passengers had to dash from one end of the ferry to the other to avoid gunshots from pirates or Mao's soldiers on the mainland, and an occasional panic among the passengers ensued whenever the boat 'tilted'.

Suddenly now five members of our family were crowding into Uncle Tai's apartment. My father and his brothers decided that we should move over to Uncle Weng's bigger apartment — the second floor of a four-storeyed shophouse — after the conversion of the wide front balcony into a room. School for me and brother Cheong seemed to become non-priority, as it was not possible for the two of us to commute daily to the schools into which we were admitted through the entrance examinations. My uncle was assured that those places would be reserved for us. Being conscious of keeping costs to a minimum, Uncle Weng and my father had to make do as a large extended family of four adults and eight children. I remember my aunt doing the daily marketing (those were pre-refrigeration days) and cooking, my mother helping with the washing and sewing, including that of the school uniforms of my cousins. The two elder girls were of my age, and the younger of the two, Yau Hung, recounted to my sisters that she had 'enlisted' me to help her in her vacation homework, which was to fill up a whole exercise book with Chinese characters — an incident that I had forgotten! I never knew whether it was rote

On The Move — An Unsettling Two Years

learning/writing or some simple punishment for mischief. In the evenings, we would sometimes stroll down the streets of the Shamsuipo neighbourhood, bustling with vendors and their wares. Sounds and sights of the lively streets were simple joys to partake of for those living in overcrowded homes. Then we were struck by a severe outbreak of measles in the heat of summer, rendering us three siblings extremely sick and bedridden for almost ten days and having to drink horribly bitter, Chinese herbal concoctions every day.

Meanwhile, communism in China had added more misery to us: my other three sisters with Aunt Yuet were temporarily 'trapped' in China, as no movement of people to HK was allowed. My father had recovered sufficiently to find work in the Military Department of Hong Kong. But the income was hardly sufficient to cater for a reasonably comfortable apartment in Kowloon and other needs of the whole family. Then an option of leaving my sibling sisters to be raised separately in China was considered. It then dawned on my parents that there was little choice but to return to Ipoh, Malaya, where the chief engineer of the PWD had left open the offer of a position to my father. By this time, a relative had brought out my sister Lai Fong from China. Finally Aunt Yuet with sisters Lai Kien and Lai Kam were reunited with us in Hong Kong a few days before we set sail for Singapore. I remember noting the sadness of my paternal grandmother as we left HK again for the South China Seas, as this was probably the last time she would set eyes on many of these grandchildren. Somehow, my memory of this return voyage by boat was very vague, as was of the subsequent, long steam-engined northward-bound train journey through the peninsula to Ipoh. Having lived through so many difficult events of sickness, of life in small crowded apartments, of separation among siblings, of seeing adults struggling to make ends meet and of dependence on close-knitted extended families, it was now a time of hope as we waited anew for whatever the future would bring. It was a joyful homecoming to the lush green environment and an abundance of bright open spaces in Malaya; so I could continue my schooling.

CHAPTER 4

Back To Normalcy

Back in Ipoh, all nine of us were housed again in two rooms in my grandmother's shophouse at 39 Theatre Street. My father's primary concern was to get us older children back to school. My brother was readily readmitted into Anderson School, my father's Alma Mater. At the CHIJ, he had to do a bit of imploring and pleading with the principal, Reverend Mother Pauline, a humble yet reverent figure from the West who had found her way to this small town Ipoh convent. She was later transferred to CHIJ at Bukit Nanas in Kuala Lumpur, and then to CHIJ in Bukit Mertajam in Province Wellesley, opposite the island of Penang, as she neared retirement. We only learnt years later from her obituary that she was a holder of the prestigious Ph.D. in Physics from the University of Chicago, and that was after I had obtained my own degree. When my father brought me to see her about re-admission to CHIJ, she insisted that I took a mathematics assessment test on the spot, which I did and she told my father the result was 'a bit weak!' However, she readmitted me into Standard 3, that is, at the level as where I should be, if without my recent school interruptions. It was already into the third (last) term of the school year, three months to final examination. It was really tough to play catch-up after having missed all the lessons of the preceding eight months, but fortunately I passed the year after all. My sister Lai Kien was to start school the following year, but there were still the three younger siblings not yet of school-going age.

We girls were lucky that our parents had selected CHIJ as our school. After a distinguished Reverend Mother Pauline with a doctorate degree, came another highly-educated Reverend Mother Paul from France, who with many others remained dedicated teachers there throughout my secondary school years. She was later transferred to Thailand, where she stayed about a decade before retiring in Nice (France), where I later visited her in Maur House (for retired nuns) in 1990. In those days, school principals played a crucial deciding role in the employment of their teachers. In CHIJ, the 'sisters' and lay teachers were all highly qualified, possessing a variety of talents — in language, play acting, music and singing, painting and art. Those were pre-Malayanisation days and many of the high-ranking officials from Britain brought with them academically qualified wives, who became our teachers. Many of them held university degrees. Expatriate wives included Mrs. Palmer, an English lady with a degree in English Literature, who with another local graduate taught us an appreciation of English classics and Shakespeare. Mrs. Scott, an Irish lady, with a Ph.D. in physics and also possessing a keen interest in biology, was in Malaya by virtue of her husband's position as Director of Agriculture, and she taught us Chemistry, Physics and Biology in Forms 4 and 5, even taking us on voluntary botany field trips in the cool evenings in the local environs. It was in such a stimulating environment where I obtained my secondary school education. There was no excuse for any academic non-performance. Indeed, from Form 1 onwards, I had performed well, not failing to top the class and the 'Form' of 250 girls in five or six classes. But I had to make a great effort in literary writing; my strength lay in Mathematics and the Sciences. For example, in Form 1, I remember scoring 98% in our first term test on Algebra and Trigonometry, while 90% of my classmates failed and a retest had to be held. I remember that in Forms 2 and 3, I found the teaching of Mathematics boringly slow, as my mind was well ahead. When I was allowed to sit in a back corner of the classroom, away from the teacher's view, I could work at my own pace, solving every problem in Algebra and Geometry in the text-books, and I believe the teacher knew that I was not paying attention, but convinced that I must be gainfully engaged in doing my own thing!

By Form 3, I was elected junior school prefect, followed by assistant Head Prefect in Form 4 and Head Prefect in Form 5. Prefects had to face school girls who could be pretty naughty in their growing teens. For example, enforcing the 'speak only English policy' in school — compulsion in the use of a foreign language — one of the jobs of prefects at recess time, was a particularly difficult duty to perform. I could not definitely say whether appearing austere produced any better result. In retrospect, I must have unknowingly developed from that time an ability for putting on an authoritative and stern countenance, without intending to be uncompromising or unkind! But overall, I had gained respect, and a following. Indeed, I was surprised when in Form 6, a classmate related that her father, an inspector of police, only gave his permission for a school excursion, after learning that I was also participating! I learnt that another parent agreed, 'if Lai Yoong goes, it should be OK!' knowing well that my parents were very cautious though not overly protective, when allowing me to take part in outdoor trips.

Life during secondary school was not uneventful as it was the time of anti-colonial communist insurgency. There were travel restrictions and sometimes night curfews and the jungle as well as surroundings were places of constant danger from gunshots from soldiers and communists. 'New villages' were constructed closer to the towns for relocating inhabitants of houses on the peripheries of jungles, so as to prevent interaction between the villagers and the communist insurgents. One memorable tragedy of national security interest was the ambush and assassination of Sir Henry L. G. Gurney, British High Commissioner to Malaya, by communist insurgents in October, 1951. The event was given wide coverage in all leading newspapers of the day. (My father had subscribed to one of the major dailies, to cater to our reading of current events). Gurney's contributions to Malaya included the many Gurney schools he had established in various States of Malaysia. His name was commemorated in the Gurney Roads in many cities, e.g. Gurney Drive in Penang and a town Pekan Gurney in Perak. A period of relative peace and prosperity (from rubber and tin industries) ensued. The second memorable event was a happy period of an extended celebration of the coronation of Queen Elizabeth II in 1953. In school, the head Art teacher, Sister Rita organized an Art Exhibition featuring coronation costumes down the ages. Being in Form 2 then, with Sister Rita as my Form teacher, I got very involved in this project. During this work, I acquired among other art techniques, a way of designing small cards, featuring greetings or verses/poems, prettily embossed with gold or silver borders.

Around this time, Ipoh (as in many other towns) was abuzz with fund-raising events for the establishment of Nantah, to be built in Singapore, (the former Nanyang University, established 1955). I remember hearsay from the adults that even the rickshaw pullers and trishaw riders were contributing their meager earnings, so keen and eager they were for a Chinese language-medium university in the region, so that students finishing secondary schooling in Chinese schools in Malaya could hope for further education on home soil.

Our extracurricular activities were enriched by the enthusiasm and innovations of our teachers. During my Form 3 year, the literature teachers (mainly Ms. Boyle and Mrs. Palmer) staged Shakespeare's *A Midsummer Night's Dream*, featuring our Form 4 and 5 girls as performers in the play. The backdrop was painted by Sister Rita and her group of staff and students. It was a lovely production that I have never forgotten. Choir productions were undertaken by Sister Anna, of Vietnamese descent; this French-speaking sister also taught us French. For sporting and social activities, students were organised into four 'houses', *viz.* GREEN, RED, YELLOW, WHITE. Indeed, girls in CHIJ were given the opportunity to develop to the fullest, without the need to go to any extra tuition classes, except for playing musical instruments which remained the preserve of rich girls.

A little about sports activities in secondary school: these were late afternoon activities, once a week. As our school did not have green space around, these were held in the fields of our sister-school, Ave Maria Convent, a school using Mandarin as medium of instruction, situated in Jalan Chung Thye Pin. It would take us approximately 20 minutes to walk there from our home. In those crime-free days it was safe for children to walk or bicycle daily to school and to sports fields. But for me, this was a much dreaded activity in the hot humid afternoons, without mentioning the irritations from mosquitoes and other tropical pests. The discomfort was aggravated by our sports attire, which consisted of starched-up white cotton blouse and blue cotton bloomers with elastic at the thigh, a prim and proper attire for young girls according to the opinion of the convent nuns of the time. Without any sports talent in my bones, such early experiences of discomforts brought on by unsuitable attire and oppressive weather, with an added misery of allergy of my feet to *natural rubber* sports shoes, resulted in my strong aversion to sports and games.

An annual inter-school function was a Form 5 debate with St. Michael's Institution (SMI). With or without much oratory skills or literary talent, I had to lead our debating team against that of SMI, led by Andrew Saw, though I am vague now about who was the eventual winner. True to his debating skills, Andrew fitted well in the law profession that he later chose. At the end of Form 5, we sat for the Cambridge School Leaving Certificate Examination (CSLCE) in eight subjects over various fields. I obtained Grade one (the top grade), but was disappointed in scoring only 7 straight A's out of a maximum of 8. Hence I have always marvelled at present day students scoring 10 to 20 A's.

Secondary School and University

Form 4 (1955) with teachers — front row (L to R): Ms. Jan, Sr. Rita, Ms. Boyle, Mrs. Palmer, Rev. Mother Paul (Principal), Mrs. Scott, Sr. Brede, Ms. Hew, Sr. Philip

Debating Teams with Rev. Brother U. Paul, Principal of SMI, 1956

As head prefect with Rev. Mother Paul and prefects, 1956

Form 6 girls, SMI, 1958

Honours class with Prof J. Miller, HKU, 1963

CHAPTER 5

Pre-University and University

The year I finished Form 5 was the second year of Form 6 education in Malayan schools. Admission was by merit for the first graders in the CSLCE. Form 6 was a two-years' prelude to university, and in Ipoh was only tenable in the three premier boys' schools, namely, Anderson School (non-religious, government, just past its centenary), Anglo-Chinese Boys' School (Methodist, government-aided, 124 years old now) or St. Michael's Institution (SMI) (LaSallian, Catholic, government-aided, centenary in 2012). It was customary for the girls from CHIJ to proceed to SMI and that was where we went, ten of us majoring in Science and a handful in Arts. The class size of 24 in Form 6 was half that of secondary school classes in CHIJ; consequently we grew to know everyone in class pretty well. I had the privilege to be Head Prefect again to represent the girls in the four Form 6 classes. Coming from CHIJ, where orderliness and cleanliness reigned supreme, where furniture and the floors shone with polishing, so much so that they could be slippery, we girls could not tolerate the worn dirty-looking desks and floor of our class room. CHIJ was privileged to have sheltered and nurtured many war orphan girls, many of whom had disabilities but were able to help in the maintenance of the school premises. In SMI, we got down to a tedious scrubbing-washing job, only to find that it was really tough dealing with an uneven floor coated with sunken-in dirt, and so we never attempted it again! Instead, we initiated other projects for the school; under the guidance of a Physics teacher — Mr. Yong Man Fong — a Physics graduate of UM, we fixed up a public address system; the boys and tomboys among us girls climbed up ladders to do the wiring job. We had other very competent graduate teachers; the Chemistry teacher was the late Mr. Lee Guan Meng, whose daughter was to become my sister-in-law. Of particular mention is a newly-arrived Brother Corkery Vincent, scholar of political thought from University College, Dublin. I remember him giving us refreshing ideas in creative writing in English Literature and in the discussion of current issues and affairs for the *General Paper*. After the launch in 1957 of the world's first artificial satellite SPUTNIK I by the USSR, he pitched our Science class in a provocative debate against our Arts counterpart, with Science proposing '*Macbeth is a greater creation than the Sputnik*', where I had the privilege to lead the team. He later served as director in the 1970s and 1980s, succeeding Brother Ultan Paul, whose 60 years on-and-off association with SMI had made him an endearing pillar of strength and encouragement to all around in SMI and Ipoh.

The institution, founded by a French Catholic missionary, is proud of its tradition in bringing up prominent leaders in society. Like CHIJ, SMI (motto: '*Ever stronger in Faith, Zeal and Community*') strives to produce well-rounded citizens in all sectors of society, for instance *via* activities like sports, the staging of annual plays (based on the classics, musicals and Shakespeare), the setting up and production of annual magazines, namely *The Michaelian*

Chronicle in 1956 and *The Explorer*, a Science magazine in 1957, of which I served as editors in 1957 and 1958, respectively. The school has among its notable alumni many prominent personalities in politics, education, religion, finance, CEO's, sports (champions in badminton, hockey and athletics) and stage and film production and celebrities. (Refer http://www.geocities.com/smi_ipoh_online/smi/smitoday/)

Let me digress to reflect a little on national affairs in 1957. That was indeed a memorable year for us all, as Malaya gained independence in a peaceful manner. It was a happy occasion with long extended celebrations throughout the country. There was much information to be gleaned from the newspapers on Malayan history and the present with new leaders and a constitutional king of the nation, the Yang Dipertuan Agong. Expectations were high that the revenue taken back to Britain by the previous colonial government could now be used to accelerate development. When the celebrations were over, however, the beginnings of racial politics were soon to consume the new country. Supposedly democratic parties mostly ended up appealing to major racial groups — a development not conducive to the harmony of the multiracial, multicultural and multi-religious environment. To be sure, politics is not a science to be contained in idealized systems or equilibrium states but is definitely too complicated to be governed by the laws of thermodynamics and quantum mechanics. This is not the place to comment on politics but it will suffice to say that my walk in the corridors of education had brought me some great friends from different racial and religious backgrounds — lifelong friends like Sushama and Zuriah (refer p. 49), and the late family friend Tun Hamdan Tahir, Director of Higher Education in the Ministry of Education, later appointed Vice-Chancellor of Universiti Sains Malaysia, and upon retirement, the Governor of Penang.

Coming back to our education, we had two options for admission into universities, *viz.* (i) *via* an entrance examination of the University of Malaya (UM, in Singapore which was part of then Malaya), to be taken near the end of the first year in Form 6, and (ii) *via* Overseas A-level examinations of Cambridge University after two years of study. The latter will provide for admission into UM (directly into Year 2 Science, on the basis of good results), and into any overseas university. Whatever our university entry preference, we were encouraged to sit for the UM entrance examination. In that examination, the first to know that I topped in pan-Malaya was my father, being congratulated by a senior mathematics teacher in Anderson School, which he regularly visited in the course of duty, being in charge of building developments in all government schools in Ipoh. However, I did not apply for admission to UM, but would rather finish A-levels and try for one of the overseas scholarships; available at that time were the Colombo Plan scholarships to Australia and New Zealand for Science, and for other fields like Geology and Agriculture. Returning scholars were bonded to contribute to the development of their home country. Having obtained straight A's at the end of Form 6, I secured a Science teaching scholarship to New Zealand, a dream for all school leavers and I was happily looking forward to going. But it was not to be, as my chest X-ray showed a shadow and I had to be kept under observation for six months, though finally that did not reveal anything health-endangering. I felt that my dream was shattered, even though UM accepted me for admission; but applications for UM scholarships were already closed, and I did not see why I should depend on my parents to finance my education, when I could qualify for any meritorious scholarship with my background of achievements. Furthernore, my father would need to think next of providing for my brother who aspired to study medicine for which there were no scholarships available. Thus I did not enter

Pre-University and University

UM but I learnt from friends that I escaped the 'ragging' or 'orientation' the senior undergraduates had planned for me, as my scholastic reputation appeared to have preceded me. However, providence was in the form of a Loke Yew Scholarship, just calling for application, for study in the University of Hong Kong (HKU). A little aside here to Loke Yew: Loke Yew (1845–1917) was a famous businessman and philanthropist during the British Malaya era. Beneficiaries of his generosity included two schools in KL, namely Victoria Institution (of which he was one of two founders) and Methodist Boys' School, Tan Tock Seng Hospital in Singapore, and HKU, which had named in his honour, *Loke Yew Hall*, wherein we students had sat for our examinations, and in 2008 HKU celebrated its Third Inauguration of Endowed Professorships with the inception of the 'Loke Yew Professorship in Pathology', established through the generosity of his grandchildren. Now, back to the Loke Yew scholarship: this was meant for needy students, but then my father was only holding a modest-paying job as a civil servant and I marginally satisfied the criteria and was successful in my application. My father and I met Loke Yew's son, Loke Wan Tho, who with a stroke of the pen wrote me a check for the air-ticket to Hong Kong — such generous members of society! Thus in the September of 1959 I entered HKU. I had been asked many times why I picked HKU; in later years I was to learn that New Zealand was at that time a great farming country, which remained for a long time the backwaters of scientific research due to its isolation from the West. Actually, although the prestigious HKU was among the best, if not actually the best, in greater Asia at that time, the fact was I did not intentionally pick it. It was the loss of a much-coveted Colombo plan scholarship to New Zealand that just providentially provided an opportunity to receive a better one, and as it turned out facilitated the link to a broader world from the perspective of chemistry, as will be evident as later events unfolded. My parents drew comfort that their first-born would have endearing relatives in Hong Kong to see to her well-being there.

    I was not alone to go to HKU but was accompanied by another Form 6 classmate, Wong Mae Lun, who was admitted for Medicine. HKU had been the choice of Malayan students for Medicine and Science after WWII till the 1960s, even after the founding in 1905 of King Edward VII College of Medicine (later UM). Mae Lun and I were roommates in Lady Ho Tung Hall (LHT) for two years, till we became hall seniors and therefore entitled to single rooms. In the halls of residence of HKU, the students were served in the tradition of the Chinese upper class where the aristocratic young ladies did not wash, cook, clean, etc. but just engaged in social pursuits. We were 'really spoilt', so to speak…we did not have to lift a finger to do even the simplest of housekeeping chores — maids made our beds and cleaned our rooms every morning including on Sundays; personal cups and glasses in our rooms were cleaned, thermos flasks filled with hot water, clothes and bedding laundered, ironed and brought back to the room. Two maids were in attendance all day on each floor which housed 20–30 girls in about 15 rooms. Those days were decades away from the arrival of mobile telephones; each floor had one phone; hence one of the main duties of the 'floor' maids was to notify us individually when we received a call, (a loud voice down the hallway in Cantonese would proclaim the person and her room). This activity was continuous all evening and night up to 10 pm, so the whole 'floor' knew who were getting calls and the length of each call! When the dinner 'gong' was sounded at 7.00 pm, we proceeded to our allocated dining table. Every year, we selected our group of eight dining mates. Every Monday dinner was 'formal', which meant that a group of ten young ladies (by rotation among all hall residents) and some members of the hall residence committee, gathered for pre-dinner refreshments in the flat of the hall warden, an academic staff. In my time at LHT, this was a Mrs. Mary Visick of the English Department. The group then entered the dining hall in a sort of ceremonial formal procession fully gowned (graduates and undergraduates in long black and knee-level green gowns, respectively) to the 'high-table' on a raised platform of about one foot height. All other diners were likewise formally attired. Announcements were then made before a western-style dinner was served. On other days, we had Chinese-style dinners. Each table was served by a maid, who would attend to all our needs, say, a second helping of rice and finally a cup of Chinese tea at the end of dinner. The high-table group finished with after-dinner drinks in the warden's flat. Such high-table dinners had followed the British Oxford-Cambridge tradition, and some 30 years later, I felt nostalgia experiencing this again in Monash University's Mannix College, where my son was residing and I was then a visiting research fellow in the Chemistry Department.

I spent four years in the University of Hong Kong to obtain a Special Honours in Chemistry. In the summer vacation after Year 1, I underwent a much-recommended surgery for a non-toxic thyroid condition, even though it was not giving me any clinical symptoms. Little did I realize that the anaesthesia I had endured had slowed down my learning ability; I found myself significantly slower in the grasping of abstract principles of Mathematics and Physics in my second year. Having been astutely sharp and quick in solutions in Advanced Mathematics in Form 6 and in the previous Year 1, I thought at the time that I had reached my intellectual limit and that Modern Algebra and Trigonometry were simply beyond my mental ability and I was even led to believe that mental capabilities peak at about age 18. I only realized years later that anaesthesia had knocked off many grey cells from my brain, after I learnt from other people's experiences and again observed similar 'brain-slowing' effects from a Caesarean section, which left me with reduced mental activity for the subsequent three months and a slow recovery spanning almost a year! When I mentioned that to my long-time friend Mae Lun, then a doctor, she tauntingly retorted 'what do you think anaesthesia does? Of course it kills some brain cells!' Later when my four-year old son needed an orthopaedic procedure to 'fix' a little dislocation at the arm, I pleaded with the orthopaedic surgeon (a Dr. Merican, husband of Ragayah Eusoff, my predecessor head prefect at CHIJ Ipoh) to have it done without general anaesthetics; in shock, he must have thought I was crazy, but assured me that the procedure would be brief.

At the University of Hong Kong we had some inspiring lecturers and we had good rapport with most of them. During those days, it was customary for university students to buy and possess textbooks; as soon as my scholarship money arrived, a large proportion would be expended in the university bookstore. Our lecture notes only served as guidelines to the texts; a lecturer like Professor J. Miller could be taking material from any section of C. K. Ingold's monumental comprehensive book entitled *Structure and Mechanism in Organic Chemistry*. We listened, took notes and then read texts; highly independent there was no need to be spoon-fed with present day *downloads, power-point* handouts and the like. We managed and performed well, unlike today's students who would complain even when notes were uploaded a bit late online. I suppose there cannot be any comparison, as unlike their counterparts of today, only a very small percentage of students of those days were able to pursue tertiary education and they were the 'crème de la crème' of students! I remember with gratitude a Mrs. Dorothy A. Collins, who had taught Organic Chemistry to Science and Medical students, amongst whom Malayan students formed the predominant group. In my conversations with her, I gathered that she must have missed that group and that I must have constituted a link to them. She had treated me with extreme kindness. On one occasion in my second year, she brought me to have dinner in her lovely apartment in a wealthy suburb of Hong Kong, and together with her husband, the Chief Government Chemist of Hong Kong (a prestigious government position), would send me back to LHT Hall after a lovely ride. We remained in touch till she passed away in Hong Kong. I learnt that she had donated substantially to the Department of Chemistry of HKU. Professor Joseph Miller, an Australian by birth, an organic chemist, assumed the Chair and Headship of Chemistry during my third year. His inaugural address was the first of such lectures I ever attended, followed soon after by that of the new Professor of Physics; I must say that I was impressed by the *nice* formality of the occasion — an entry procession of the new professor with the Vice-Chancellor, the Dean, and other senior academics in flowing academic gowns, the Governor as guest of honour with other dignitaries in the front row, the lecture theatre filled to overflowing. That was an elegant legacy of the tradition of British universities, an occasion to *show off* and introduce their new professors to the public. For us Honours students, he immediately initiated research projects. There were only thirteen of us in the class. I selected one of his projects on 'Phenylation of Azobenzene' and had my first taste of the demands of research, like doing tedious gas chromatography separations well into midnight using the first generation instruments, which did not always work well. However, living in a hall of residence at just a stone's throw away did facilitate such odd hours of work. I believe that efforts in any endeavour always have good outcomes, and mine was a contribution to a paper published a few years later in *The Journal of the Chemical Society*.

As a former researcher of University College London (UCL), Professor Miller knew Professor Ronald S. Nyholm and so had invited and appointed him to serve as our External Examiner for the Honours degree. On hindsight, I realized it was this choice of J. Miller that influenced the course of my career and later life. Professor Nyholm brought with him the enthusiasm surrounding the new world of Coordination Chemistry, which contrasted well with the classical facts-full descriptive Inorganic Chemistry, which we had been taught but rarely appreciated. In a small department with only a handful of graduate students, we Honours students were scheduled to attend all seminars in the department. In fact, these took priority over any overlapping lectures. Nyholm's tales of exquisite ligands, based on P and As, presented in two seminars, convinced me about my future research directions. Thus a dream had begun. In HKU Chemistry Department, we were kept aware of the career advancement of most of our seniors; in particular, those who had gone overseas for Ph.D. degrees and the returning ones, e.g. Chan Sai Cheong returning from UCL was among our role models, though few of the Honours graduates in my time felt motivated enough to follow in their footsteps. In my case, after obtaining an Honours degree, Professor Miller could not persuade me to stay behind for a Ph.D. but I chose to return home to Malaysia to apply for a Commonwealth Scholarship to UK for a Ph.D. in UCL. Thus, I bade farewell to HKU. Professor Miller later left for the University of São Paolo in Brazil, and while there he had visited us in UM on two occasions in the 1980s, during his travels back to East Asia.

My stay in Hong Kong had been more enriching than from just a formal educational standpoint. It had given me an opportunity to re-foster family ties with my grandmothers, uncles and aunts and my cousins, who had earlier helped us in the critical phase of my father's surgeries and hospitalization. There was also the chance to get to know the younger cousins born after 1950. Regular weekend

Pre-University and University

dinners were with Uncle Tai's family, which now included his children, Rosa and George, in his apartment in Hing Hon Road, within walking distance of LHT Hall, and occasional visits to my other uncles — Uncle Chung Wing in Shamsuipo and my maternal Uncle Chee Hing in Kowloon. I had observed with dismay the different strata of society in Hong Kong — the rich living on the hills on the island, the upper middle class in suburbs like Kowloon Tong and the poor living in government 'stone houses' in an area established after a great fire gutted their previous slum dwellings. I had been dazzled by the lavish displays in departmental stores on most occasions as a window-shopper, whilst numerous stalls along the narrow meandering streets catered to the essential necessities of the poorer citizens. I had tutored a Form 2 student in Science and Mathematics during my Year 2, considered a 'honeymoon year' because there were no examinations. I had made many friends among class and hall mates, and members of the Malayan and Singaporean Students Association, now scattered around the globe. For instance, just mentioning a few, I still maintain contact with Lily Seet (Singapore), Joyce and Yu Fai (Toronto), Poh Yu (Vancouver), Effie Chang (New Haven, Connecticut) and Athena Chow (Melbourne).

Equipped with a degree, would it be possible to blaze new trails? In those days, in the view of relatives and friends, a degree was perceived as a necessary ticket to good jobs — a professional degree to the designated profession and a general degree in Arts or Science mainly to teaching. In Malaysia where I planned to return, there were some opportunities available for Science graduates in a few research institutions, namely, The Geological Survey and The Department of Mines, both in Ipoh my hometown, and The Rubber Research Institute in KL. Postgraduate studies had become attractive as a work-*cum*-study option. Besides I could be financially independent and now perhaps provide a little to my parents who were still burdened with support of my younger siblings. However, to take a big plunge into research is an adventure in life as it meant venturing into the bigger world, as the small world in local research institutions may not have enough positions for higher degree holders and while the academia will undoubtedly be looking out for the world's best among the highly educated elite, even in those days. With an interest in the 'unknowns' in chemistry enkindled during the degree course and intensified in the last stage by an external examiner, I nevertheless decided to take the gamble, but first return to Malaysia to get a scholarship from home soil.

CHAPTER 6

In The University of Malaya, Kuala Lumpur

There were benefits to be back in KL for an interim period. In a way, I was homesick and was keen to be home or nearer home in Ipoh. In those days, communication was *via* letters or telegrams for urgent matters. I wrote weekly or biweekly letters to my father who always replied promptly. Imagine me describing to him, an engineer, my first observation of beautiful crystals in organic chemistry laboratory in my second Year! My command of the Chinese language was not sufficient for me to write to my mother in Chinese, though I had had home tuition in Chinese by a private tutor. I regretted then that I had not gone with my brother Cheong to the night Chinese School of Ho Tih Ann (my mother's teacher), which was just opposite 39 Theatre Street where we had lived for about five years. Cheong could write to my mother in Chinese. I hardly wrote to my siblings, knowing that my father would pass on my news to them. Thus I was slightly out of touch with them. Brother Cheong had started studying Medicine in the University of Malaya in Singapore. Sister Lai Kien had proceeded to Trinity College Dublin for a course in physiotherapy.

Sisters Lai Fong and Lai Kam were already in upper secondary school in the CHIJ, all performing well, in the 'legacy of the eldest Wong sister!' so they said. The youngest brother Kai Fatt was in secondary school in Anderson. Previously I had been home for two summer vacations, travelling by ship, as customary in those days of relatively expensive air-travel. Contrary to our earlier dreary voyages of 1949–1950 to Hong Kong, these two sea-trips were actually enjoyable, being berthed in cabins shared with Malayan university girl friends. It was like the beginning of a small cruise liner in its days in infancy. On one occasion, the ship stopped over in Manila, and we had a tour of the city, guided by two Filipino fellow passengers.

The application for any overseas scholarship would take at least a year to process. Professor J. Miller had advised me to see Professor Rayson L. Huang, his friend in the Department of Chemistry (UM in Kuala Lumpur) for a research opportunity. The University of Malaya was established in 1949 in British Malaya by the merging of King Edward VII College of Medicine (founded in 1905) and Raffles College (founded 1929). Like all universities of the British tradition of that time, the institution was comprised of Faculties each under a Dean, and each faculty comprising of departments under a head who was the one and only professor in the department. In 1959, the University of Malaya established an autonomous branch in Kuala Lumpur, with faculties in Arts, Science, Engineering and Medicine. Rayson L. Huang was the founding professor in Chemistry. He offered me an appointment as a tutor, which carried with it a stipend of RM600 a month, with encumbent duties of 12–18 hours of laboratory demonstration per week during term time plus additional marking of laboratory reports. The main role of a tutor was concurrent research work towards a higher degree, beginning with a master's and extendable to a Ph.D. My preferred area of research was coordination chemistry, but there being no

In The University of Malaya, Kuala Lumpur

staff in that area of research at that time, I was offered a project in free radical chemistry, under the joint guidance of Professor Huang and Dr. Lee Hiok Huang, an MIT Ph.D. who had moved from the campus in Singapore to be part of the growing department in Kuala Lumpur. I was very fortunate, for I received excellent training in both research methodology and scientific writing under Dr. Lee in the tradition of MIT. I finished a master's thesis in seventeen months, during which time I made several life-long friends/associates, amongst whom were to be my future husband Goh Swee Hock (a top scholar of his time, one of the first batch of students at UM, the first Malayan to proceed to the University of Chicago for a Ph.D. under a Fulbright scholarship and set the precedence for two other First Class Honours from UM to follow, *viz*. Lim Poh Eng, now in Universiti Sains Malaysia and Lee Soo Ying, former Dean of Science in NUS and founding Dean of College of Science in NTU), Tan Chee Hong, a successful entrepreneur (Matrix Sdn. Bhd, Matrix Group of Companies), Mohinder S. Malhotra, JP, retired soil scientist in RRIM, and one-time Director of Science (presently director of cooperative groups, and of a school for disadvantaged students), and Esah Sieh (RRIM, presently Director of the Malaysian Rubber Export Promotion Council, in Washington, DC). Research in KL was pleasant, although lacking in academic vibrancy, as well as, in the latest instrumentation. However, I managed to get sufficient results for a paper published in *The Journal of the Chemical Society* — my first publication. The group of us had developed the research culture of hard work. My peers in the laboratory were a jovial lot. One unusual recollection was of Mohinder dashing into the laboratory one morning and announcing that President Kennedy had been assassinated, but we did not respond in shock, not knowing whether it was a joke or sober reality!

Towards a serious post-graduate education, I was successful in my application for the Commonwealth Scholarship, in which I had stipulated Chemistry Department of University College London, as the choice for my place of study. It was the beginning of a dream come true to be offered a place of my selection, but I was reminded of my parents' anxiety. In those early days of an independent nation of Malaysia, it was not usual for anyone, especially a girl, to be pursuing higher degrees. I already had obtained a Master's degree; surely I should be settling down. Our family friends, even from among the educated circle of teachers and headmasters, wondered why my parents would allow a girl to proceed further. I remembered promising my mother that after the Ph.D., there would be 'no more degrees'! Having a bachelor's and master's degrees were fine but one thing was certain — the intellectual challenge was still missing. I already knew quite rightly my science would remain limited and views myopic without greater exposure to the environment of great scientists and their centres of excellence. Thus it would be the lost opportunity of a lifetime not to take up the offer to be immersed in a place of learning where the changing frontiers of science are being pressed forward.

In The University of Malaya, Kuala Lumpur

In Laboratory

With Joyce Lee, Science Exhibition
HKU, 1961

With Esah Sieh, UM, 1964

In lab. D18, Chemistry Dept,
UCL, 1966

With Richard Wong, UM, early 1990s.
In the background, a VAC Drybox,
the first in academic institutions
in Malaysia

CHAPTER 7

Postgraduate Studies in London

I arrived in Heathrow on January 5, 1965 on a bleak wintry morning, took an airport bus to Victoria Station where by previous arrangement I was met by Wong Ka Wing, a girl two years my senior in HKU, then pursuing a Ph.D. in Organic Chemistry in Bedford College, London. We went to Chelsea to get me checked into Crosby Hall, a residential hall for university women (British and Commonwealth). The Hall was opened by Queen Mary in 1927 and I had found out that it was sold in 1992. The premises were old but imposing, especially by the dark long but wide hall with a huge high-table at one end in front of a large mural. It appeared to resemble a hall that I had seen in films of Henry the Eighth and it had struck me as eerie. This served as the dining hall. We next went to Oxford Street to shop for my settling-in requirements. I was tired out by jet-lag and my shoulders were weighed down by a heavy winter coat. I felt miserable seeing the unfamiliar gloomy skies in the early hours of the afternoon and later losing my way on the return to Crosby Hall, despite holding and referring to the 'London A to Z' travel guide book! The following day saw me making my way by bus and the Underground to University College in Gower Street. The commuting took more than an hour — which I considered a waste of precious time! As chance would have it, I met in the large city of London, Ragayah Eusoff, who introduced me to Ms. P. C. Sushama, a lecturer in Social and Preventive Medicine of the teaching hospital of UM, spending a one-year's stint in University of London teaching hospitals, on a study-survey on the interaction between social workers and medical students. She was living in a double room by herself in Lillian Penson Hall (LP) in Talbot Square near Paddington Station. This is a hotel-turned postgraduate hall of residence of London University, and today it comprises some 300 units of single and double rooms and small flats. The renovated hall was bright and modern, with central heating. It was a far cry from the chilly and gloomy atmosphere I experienced in Crosby Hall. Relatively it was luxurious with attached bathroom, completely serviced like a hotel. More importantly a direct ride on the Underground, from Paddington to Euston Square on the Circle Line, would get me to UCL Chemistry Department in about 30 minutes. Unfortunately there was no vacancy. I therefore pleaded with Sushama to allow me to share her double room, and as it turned out, we had an enjoyable one-year's room-sharing till her return to UM. Together we had enjoyed the rich cultural activities of London — concerts, operas, plays and ballets, on the cheapest tickets available. After Sushama left, I attended ballets with another girl friend on 5-shillings tickets in Covent Garden, in the high 'standing' space at the back of the theatre. Weekends were partly for supermarket shopping. On weekdays, Sushama would often have evening meals ready when I returned from the laboratory, that is after we moved to another room with an attached cupboard-like stove, like a standing 'fume cupboard'! Here I stayed till the end of my candidature in UCL, upgrading slowly to a flat with a *real* kitchenette, sharing with Chan Lian Hong, a First Class Honours girl from UM, doing an M. Phil. in Analytical Chemistry at IC. My last

Postgraduate Studies in London

hall-mate was Marina Majumder (medical doctor), and through her I met Zuriah Sharifah Aljeffri, a Malay girl who was to develop into a successful artist in Chinese water paintings (www.zuriahaljeffri.com). LP Hall was very cosmopolitan, with graduate scholars from many Commonwealth countries. Memorable were the days in 1965 after the unilateral Declaration of Independence from Britain by Rhodesia (later Zimbabwe) — the African scholars, mainly from Nigeria, the Sudan, Rhodesia and South Africa, were visibly very upset, even angry, whenever the TV newscaster commented on the 'political immaturity' of the native Rhodesians.

I was to learn later that LP Hall was the hotel made notoriously infamous by the Profumo Affair, a political scandal (1963) in which then Secretary of State for War, John Profumo, was involved in a brief relationship with Christine Keeler, a showgirl who was at the same time the mistress of a known Russian spy. The scandal forced Profumo to resign and severely damaged the reputation of Harold Macmillan's government.

Continuing to describe my second day in London: upon arrival in UCL, I made my way to the Department of Chemistry to meet Professor Nyholm (HOD). I learnt later that Nyholm already had a large research group and was also very involved in educational committees in London. So he assigned me to a project of a young but dynamic staff member, Michael D. Johnson with a more senior co-supervisor, Martin L. Tobe, a Reader in the Department. I was mildly disappointed, because I had selected UCL for Nyholm's type of coordination chemistry. However, it was not the end of a dream, and I settled down to work diligently. At that time, I still had not realized that I had unwittingly selected a very illustrious institution, founded in 1826, as London University, by the Scotsmen James Mill and Henry Brougham, under inspirational guidance of the philosopher Jeremy Bentham. The third oldest English university after Cambridge and Oxford, UCL had a noble beginning, being the first British university established on a secular basis and also the first to admit students regardless of race, class, religion and gender, hence the first to admit women. It had consistently ranked among the best universities in the world, with an emphasis on innovative teaching and research excellence. It had nurtured scholars in varied areas of study, and obtained its 20th Nobel prize in 2007. Its eminent alumni include world leaders, e.g., M. Gandhi; prime ministers, presidents and founding fathers of many nations; chief justices, attorneys-general, a Lord Chancellor and a Chancellor of the Exchequer, a judge of the International Court of Justice at the Hague; prominent poets, e.g., Robert Browning; famous scientists, e.g. Alexander Graham Bell and Francis Crick; and many leading journalists, designers, media personalities and businessmen.

The department in which I found myself was likewise one of much acclaim and tradition, with perhaps a rival from Imperial College (IC). Shortly after I arrived in UCL, G. Wilkinson from IC reported his new rhodium hydrogenation catalyst (*JCS. Chem. Commun.*, 1965, 131). Soon after, Nyholm hosted him for a department seminar, and I could sense a friendly rapport, despite a hearsay rivalry, between the two departments. UCL Chemistry possesses among its notables, the Nobel laureates, Sir William Ramsay (who shared the prize with Lord Rayleigh of Cambridge in 1904, for the discovery of the inert gaseous elements in air), and Sir Robert Robinson (1947 for natural products), who was Professor of Organic Chemistry (1928–1930) at UCL. The centenary anniversary for Ramsay's Nobel prize was celebrated in December 2004, when the Royal Society of Chemistry, nominated UCL as a Historic Chemical Landmark site.

In 1965, there were two resident knights of the realm in the department, namely Sir Christopher Ingold and Dame Kathleen Lonsdale. It was in UCL that Ingold carried out work which rationalized Organic Chemistry on a mechanistic basis, reported almost exclusively in the *Journal of the Chemical Society*, and in the two editions of his book *Structure and Mechanism in Organic Chemistry*. Today his contributions are remembered in a Chemistry building in Gordon Street, named in his honour and in a commemorative plaque at the entrance, bearing the citation *'During the period 1930–1970 Professor Sir Christopher Ingold pioneered our understanding of the electronic basis of structure, mechanism, and reactivity in organic chemistry, which is fundamental to modern-day chemistry.'* Dame Kathleen Lonsdale (née Yardley) was a crystallographer, who established the structure of benzene by X-ray diffraction methods (1929) and hexachlorobenzene by Fourier spectral methods (1931). She was conferred a D.Sc. of UCL (1936), while working with Sir William H. Bragg (1915 Nobel laureate in Physics) at the Royal Institution. A mother of three children, she was a female scientist of many *firsts*, including first woman tenured professor at UCL and head of the Department of Crystallography (1949), first woman elected Fellow of the Royal Society, first woman president of the International Union of Crystallography, and first woman president of the British Association for the Advancement of Science. She was made a Dame Commander of the Order of the British Empire in 1956. A Kathleen Lonsdale Building is named in her honour today at UCL. *Lonsdaleite* an allotrope of carbon, a rare form of diamond found in meteorites, was named in her honour, a tribute to a pioneer in the use of X-rays to study crystals, in particular of diamond. Indeed, the name and greatness of an institution is entwined with that of its faculty members, past and present.

I soon found that I was one of some 150 graduate students and post-doctoral fellows in the department, from numerous Commonwealth countries and the Americas. Among these researchers, the Australians predominated, so much so that my fellow peers had talked of Australians *versus* the rest in football matches! Nyholm, of Australian origin, coming from Golden Hill in New South Wales, was the magnet for good scientists to UCL. A Ph.D. student of Ingold, he effectively initiated the 'Renaissance of Inorganic Chemistry', on which he published a key paper (1955). He was developing Coordination Chemistry with new phosphine and arsine ligands with transition metals in varied oxidation states, in a direction which had invigorated Inorganic Chemistry, in fact, which had attracted me from HKU, and which had no parallel in N. America, according to the American and Canadian post-doctoral fellows in UCL at the time. Today undergraduate students will remember him for the VSEPR (*valence shell electron pair repulsion*) theory, which he developed with R. J. Gillespie (then lecturer at UCL, now Professor Emeritus at McMaster University in Canada). Nyholm was knighted in 1967 and it was obvious that his research would soon be extending to cover the whole Periodic Table, but this was curtailed by his untimely sudden death in a car accident in 1971. His monumental interest in Chemistry was reflected in his activities and publications; more importantly he left a great impression on me as well as others to pursue our goals in chemistry. At UCL, he sought to have Chemistry treated as a unified discipline without the artificial divisions into inorganic, organic, physical, etc. In line with this philosophy, graduate students working in different areas of research were all intermingled, a new arrival being allocated wherever there happened to be an unoccupied bench.

Thus I found myself given a bench in laboratory D18, a large laboratory with six double-sided 8-feet long benches, stretching to the side wall containing the windows. On the opposite wall were situated four large fume-hoods; two more were situated at both ends of the laboratory. The laboratory could accommodate twelve researchers, amongst

Postgraduate Studies in London

whom were three post-doctoral fellows (Australian, American and Canadian), and Ph.D. students from Jamaica, Nigeria and Sri Lanka, in various chemical sub-disciplines. Level D in the building was the basement; the floor levels go up to C, B and A, and above that was erected Level Z, in one section of the building. Looking up at the high window from my bench, we could see pedestrians walking by on the pavement of Gower Place, and occasionally sunlight streamed in on bright summer days; on such rare occasions, my Jamaican friend, Oliver, on my opposite bench, and most others as well could not resist the temptation of a walk in the park, postponing or putting on hold all work! In the building, there was also a penthouse which housed permanently a safety superintendent, whose job was to check the log-in book at the entrances every two hours during after-hours (every night and weekends), followed by a walk round the building, in particular to the locations indicated by the researchers in the entry log-in book. It was also his job to check at 10 pm on all overnight and weekend experiments, and 'shut down' those not carrying a description of the reaction. At one end of laboratory D18 was the side-entrance to the building from Gower Place, close to Euston Square; the alternative entrance to the building on Level C was through the main entrance of the College, past the South Cloister where was displayed the preserved corpse of Jeremy Bentham, the spiritual father of UCL. (See 'Recollections' p. 57). Apart from the confusing design of entrances to various laboratories, there was the important close access to chemical literature. Close by the Chemistry entrance was the UCL Science library, which housed all the Chemistry journals and books. We also enjoyed the luxury of a departmental library, which had a collection of the major chemistry journals. The proximity of the two libraries to our laboratories facilitated reading of chemistry literature, which we could sometimes even do concurrently with bench work. This was important as the age of photocopying had not arrived, not to mention the personal computer, ubiquitous nowadays, but still more than two decades away.

My project dealt with studies on σ-bonded organometal complexes. My initial attempts to resolve a chiral pyridylalkyl bromide (for further reaction with chromous ions) took a year to materialise — some of the dreamed-up designs of chemical reactions did not see the light of day. Hence it was the concurrent work on σ-bonded organocobalt complexes, a distant 'cousin' and the simplest analogue of vitamin B12 coenzyme, that progressed much further ahead. Mechanistic investigations of their reactions under acid conditions required kinetic runs which often spanned a good 10–12 hours each. This meant long hours each day in the laboratory, and when reactions proceeded slowly, they continued to 11 pm, as happened on several occasions. On one such night, Mike Johnson came by and said he would give up chemistry if anything happened to me. What a way to impress on me the need to exercise caution, as my kinetic reactions were progressing to high concentrations of perchloric acid! In another aspect, there definitely was no lack of physical exercise in these kinetic runs, as frequently the experiments would involve the use of thermostatic baths and spectrophotometers housed on various floors of the building, including Z-floor. Satisfaction was when my long hours paid off; by Easter of 1965, we (actually Mike Johnson mostly!) had figured out the mechanism of the reaction and after two and a half years, I had obtained sufficient results for the thesis. At that time, my work had resulted in three research papers.*
But at that point of time, it may still have been questionable whether I had matured sufficiently for independent academic research. Hopefully, I had, after having been imbued all this while in the vibrant intellectual environment

---

* my early papers listed my name as *Lai Yoong Wong*.

of UCL Chemistry Department, which provided for weekly Inorganic lunch meetings across all groups, weekly departmental colloquia by Ph.D. graduating students, staff and visitors, as well as voluntary attendance at selected lectures and courses for graduate students. I therefore consulted my two supervisors about writing up. Professor Tobe asked me about future plans; upon learning of my intention to go to Chicago, he said he could recommend me to his friend Jack Halpern, already one of the named distinguished professors. Thesis writing began in earnest. When it was done and it was time for Dr. Johnson to review, he suggested doing it in a more comfortable and relaxed atmosphere in his home and he would drive me to North London and go over the thesis with me over tea prepared by his wife Fiona, and we carried on to supper, after which he drove me back to LP Hall. There were two such occasions. What a gracious supervisor I had! So I got to know his young family very well. Later, while at NUS, around 1999, I met Fiona who had come to visit their eldest son Christopher in Singapore, then Reuters' representative. Over dinner, Christopher reminded me that I was the one who taught him how to handle the chopsticks when he was seven years old! (something that I had forgotten).

My thesis was typed by a new department secretary, and copies made on a first generation photocopy machine, which unfortunately used photo-degradable ink! Thesis completed and presentation slides ready, being prepared by the photography section of the department, it was time for me to give my graduation colloquium. These were attended by most of the department; normally present were the distinguished professors, Professor Sir Christopher Ingold (Organic), Professor Sir Ronald Nyholm (Inorganic) and Professor D. P. Craig (Physical). The last named chaired my session, and I had opportunities to meet him again on his visit to UM (*ca.* 1970), and in later years during my visiting fellowships at the RSC of ANU, where he relocated in the 1970s. The one-to-one oral examination with the external examiner, Professor M. Green, well-known in organometallic circles, from the University of York, followed on another day. Naturally, I was nervous, but I remember that Professor Green was very kind and prompted me when I 'stumbled'. There was little need for any amendment. That evening Professor Tobe and his wife Rosalie graciously hosted a celebration dinner in their home for Professor Green, Dr. Johnson and Fiona and me. A few days later, I left London, sent off to Heathrow Airport by Dr. Johnson. Almost two months later in Chicago, I received by mail the prestigious William Ramsay Memorial Gold Medal, awarded to the most deserving Ph.D., presumably based on the thesis and the colloquium. Imagine my surprise, as frankly I was not even aware of the existence of such a medal. It was most unexpected, almost like the continuation of a dream, but a humbling experience. I could hardly believe my fortune to be selected from amongst equally deserving candidates such as Poon Chung Kwong. It was later when I read more about William Ramsay that I believe that my qualification as a member of the female minority must have added weight in my favour, as William Ramsay, together with Augustus Harcourt and William Tilden, had fought arduously for women chemists' rights, including admission into The Royal Institute of Chemistry and The Chemical Society.

Thus unwittingly, I had drifted into and obtained a Ph.D. from an eminent department in an illustrious college. Indeed it was a rich environment where I spent thirty-two months, doing research in elucidating mechanisms in organometallic chemistry, attending so many educational seminars, discussing with peers, and in the process forming long-standing professional links with staff and postdoctoral fellows, who were to remain as my international colleagues in academia — notably, then lecturer Martin A. Bennett, (FRS, Professor, RSC, ANU, Australia), with whom I spent many a sabbatical and research leave in later years;

Postgraduate Studies in London

Recollections

Jeremy Bentham's preserved corpse, dressed in his own clothes, displayed in UCL's South Cloister[a]

First Controlled Nuclear Chain Reaction Monument, U. Chicago[b]

Sir William Ramsay[d] and his memorial gold medal awarded to Wong Lai Yoong, 1967

R. S. Nyholm[c]
(HOD, UCL Chemistry Department)

M. L. Tobe[c]
(Ph.D supervisor)

D. A. Collins[e]
(a dedicated lecturer, HKU)

[a]Source: Per kindness of M. D. Johnson, UCL Chemistry Department
[b]Source: Per kindness of S. H. Goh
[c]Source: With permission. From UCL Chemistry website — A Periodic Table of Lecturers
[d]Source: With permission. From the American Institute of Physics
[e]Source: HKU Chemistry Department newsletter

ICI postdoctoral fellow Glen Deacon, (Professor, Monash University) with whom I have been keeping up research collaborative links, in particular under a grant from Australian Department of Industry and Technology, Bilateral Science and Technology Collaboration programme for work on Manganese Chemistry at Monash University, which allowed for several short visits in 1993–1994, and again after my retirement from UM, on a University-Industry research project on '*Electric Windows*' based on ruthenium dyes for solar cells. Then there was B. Bosnich, an honorary research fellow, who often walked through my laboratory D18 on many a morning with greetings to brighten up the day, and who soon after took up the Chemistry chair in the University of Toronto, followed by The University of Chicago. Also working in D18 were Robert W. Cattrall (now Emeritus Professor, La Trobe), John Johnson, postdoctoral in Physical Chemistry from USA, Nithy from Sri Lanka, (Ph.D. student of M. L. Tobe), who 'rocketed' a thermometer to the ceiling while doing a distillation in a closed setup; and Ali Hussein (Ph.D. in Physical Chemistry) from Sudan, and in an adjoining office was friendly crystallographer Peter Pauling, son of Linus Pauling. Working on an upper floor, in Tobe's group, was Poon Chung Kwong, one year my junior in HKU, but ahead of me in UCL, having started his candidature immediately after his Honours. He proceeded to an academic career in the University of Hong Kong after graduation and later became the president of Hong Kong Polytechnic University, from which post he has just retired.

Sometime during my first year in UCL, Professor Rayson L. Huang came to visit Sir Christopher Ingold, with whom he was associated while at UM in Singapore. He also took the occasion to introduce me to Sir Christopher. I was also very happy to meet Dr. H. H. Lee and his family when he returned after a sabbatical leave in Stanford University *via* London. I have kept close contact with my teachers and

Postgraduate Studies in London

supervisors from whom I have learnt so much. We still get together in Singapore whenever Rayson Huang passed through en-route to Hong Kong from Birmingham where he now resides (see 'Ph.D. and post Ph.D.', p. 123).

London with a population of 7–8 million in the 1960s has always been a bustling city, with crowds of people always on the move. Together with crowds of fellow commuters emerging out of the Underground at Euston Square each morning at rush hour, I had often imagined myself being part of an army of rats rushing out (escaping) from underground into the upper external road. Yet in all this daily helter skelter a familiar face appeared in one of the usual gloomy mornings. What a surprise to recognize among the crowd a former classmate's brother Francis Lip, unknown to me to be in London at the time. He was studying architecture and accompanying him was his fiancée Evelyn, both going to college. After obtaining the Architecture degree, followed by a short period of teaching in a polytechnic in Selangor, Evelyn relocated to the School of Building and Design in NUS where she served till retirement in 2000. She is among the few friends I can keep in touch with in busy Singapore; unlike me she is a lady of many talents — artist, author of Feng Shui books, and post-retirement a Chinese operatic singer and performer.

During my first Easter break in 1965, I made it a point to visit my sister Lai Kien, in the second year of a Physiotherapy course in Dublin. In those days of barely affordable airfares to UK, like many students of her time, she would not get to return home till completion of her course. Still very wintry, April was not a good time of year to visit Ireland. Despite that, I had travelled north to visit former Science teacher Mrs. Scott and family, living in a suburb of Belfast. They treated me to my first taste of salmon, having kept frozen one from a previous fishing trip, for my impending visit, knowing that salmon fish was not available in then Malaya. There was no talk of being/behaving 'green' in those days, but I learnt that a metallic bin kept outdoors was as good a freezer as any during wintry nights in that part of the world. After this, I saw Mrs. Scott again in PJ in 1974 when she returned to Malaysia for a visit and graced our humble terraced house as guest, before moving on to the luxurious bungalow of her good friends, the Suffians, Tun Mohamed Suffian being the Lord President (Chief Justice) of Malaysia (1974–1982) and also Pro-Chancellor of UM.

There were many advantages of being scholars in a foreign host country. We were invited to partake of the Christmas and Easter trips arranged by the British Council to various parts of the country, e.g. to Shakespeare country, in particular Stratford-upon-Avon, to Coventry in historic Warwickshire, to Scottish Aberdeen, to a stay in a stately home in the Midlands, and to dinner in the luxurious home of one of London's distinguished citizens in Kensington. I had the occasion to visit historic and beautiful Stratford-on-Avon for a second time, this time with my sister Lai Kien (from Dublin), Brother U. Paul (from SMI, Ipoh) and Tan Chee Hong (from Welwyn Garden City), when they happened to visit London at the same time. Further, in order to experience more of the country, I had participated in chemistry conferences of *The Chemical Society* (London) in Sheffield and in Aberystwyth, a lovely city on the west coast in Wales. It was tempting also to see part of Europe, just across the Channel, so near and yet so far! Unbelievable as it may seem now, with a monthly scholarship stipend of 56 pounds sterling, later raised to 80, I could save enough for two short holiday trips with my sister Lai Kien, in 1965–66 — one to Paris to walk down Champs-Elysées Avenue, to visit the Eiffel Tower, Notre Dame Cathedral and Versailles Palace, and the other to the Scandinavian countries to witness the midnight sun and sail into the spectacular Norwegian fjords. The latter experience was

Postgraduate Studies in London

captivating. It was on occasions like these that one could get refreshed and rejuvenated from the sometimes gruelling laboratory experiments. Science and innovative ideas can bloom when the imagination runs wild while watching majestic fjords or just keeping still staring at the wide expanse of waters or the steep inhospitable cliffs. It was in such admirable surroundings where insurmountable experimental problems somehow could dissolve in the sparkling blue waters and where barriers could be challenged by waves of thought-provoking ideas. My supervisors were only too happy that I took time off to refresh and dream anew, though leave was not stipulated under the terms and conditions of my scholarship. In my opinion, that was rightly so, as a scholarship is an award, in fact a privilege, and not a job appointment.

CHAPTER 8

Beyond The Ph.D. and Chicago

The flight from Heathrow took me to Montreal where I met Swee Hock (SH) to spend a week at the first World Trade Exhibition (EXPO '67). It was a hectic few days in the summer heat in the crowded EXPO grounds where a huge beautiful geodesic buckminsterfullerene-like structure (see back cover page) was highly conspicuous even before buckminsterfullerene (C_{60}), named after R. Buckminster Fuller (1897–1985) was to be discovered (1985). Rational synthesis of such a large molecule remained beyond presently available methodologies although cubane (C_8H_8) had been around (Chicago, 1965) while tetrahedrane (C_4H_4) derivatives and dodecahedrane ($C_{20}H_{20}$) were later synthesized. Accommodated in a youth hostel, we travelled daily in the subway train, which was remarkably clean, being newly completed in time for the EXPO. As we walked the streets, we also noticed the relaxed living style of the Montrealites, sitting on their front verandahs to watch the world go by, understandably also to get reprieve from the summer heat, as many were manually fanning themselves.

A leisurely southwesterly train journey to Chicago introduced me to some scenic views of that part of the continent. With an offer of a position as research associate by

Professor Jack Halpern, I arrived in the prestigious University of Chicago (UC), where the many old but distinguished buildings had been home or school to so many Nobel laureates. Halpern's laboratories were sited in the newer Searle building, adjoining the Noyles building, both possessing very impressive facades and rich architectural character in the tradition of much of downtown Chicago. The group size was about ten, half of whom were post-doctoral research associates. The main focus of research in the group was mechanistic studies of oxidative addition, hydrogenation, and other catalytic processes, coupled with some syntheses. Much of the work would invariably involve kinetics, under nitrogen in aqueous systems. I was exposed to fast kinetic techniques, of which I only used 'stopped flow' on occasions. At that time, SH was in his final year of Ph.D. research in the group of G. L. Closs, noted for carbenes chemistry and later the NMR 'CIDNAP' phenomenon of radical pairs. Himself living in International House, SH had booked for me a small furnished, partly serviced unit in Blackstone Apartments, at the corner of Blackstone Avenue and East 55th Street, ten blocks from the Chemistry Department. The twenty minutes' walk home at night was a daily risk in the tough neighbourhood around Hyde Park which was surrounded by depressed inner city areas. Fortunately, I did not encounter any untoward incident.

On a clear snowy day, a stroll across the old Staag field would bring one within sight of the Nuclear-Reaction sculptured monument (see 'Recollections', p. 57) amidst the surrounding whiteness as a grim reminder of the power of atomic fission. History (1942) was made in UC that would change the world forever when a group of scientists with E. Fermi unleashed a sustainable nuclear chain reaction that made available nuclear power for peaceful uses, despite the first application as a uranium atomic bomb. Soon afterwards, in the humble, relatively decrepit Jones laboratories Glen T. Seaborg's group went on to isolate plutonium also

Beyond The Ph.D. and Chicago

for a fission bomb. In 1949 Willard Libby developed radiocarbon dating. Harold Urey, who had earlier discovered deuterium isotope and heavy water in 1931, conducted the famous 'Miller-Urey experiment' on the origin of life's amino acids in 1953.

The string of Nobel laureates associated with the University of Chicago continued unabated among scientists, and later economists, over the years in many areas. Indeed, the number of luminaries is comparable to that of Cambridge and exceeds those of universities of the rest of the world. When I was there in 1967, the President was the Nobel laureate G. W. Beadle (Genetics, 1958). Now and then, the campus was filled with excitement; in chemistry an atmosphere of grandeur still pervaded through the laboratories from the previous year's announcement of Nobel laureate R. S. Mulliken for molecular orbitals. The lecture hall in which this grand old chemist gave his 'sit-down' talk/address to the Chicago community after the Nobel award was packed to overflowing. Then there was talk that Y. T. Lee (later to be the 1986 Nobel Laureate) had been offered employment in the faculty. A number of other Chemistry laureates had already been associated in some way with UC, including W. F. Libby (1960), K. Bloch (1964) and K. Ziegler (1963) and later G. Herzberg (1971), W. H. Stein (1972), I. Prigogine (1977), H. C. Brown (1979), H. Taube (1983), F. S. Rowland (1995), P. J. Crutzen (1995), R. S. Smalley (1996) and I. Rose (2004). The free academic environment attracted many of the best scholars especially in Physics and other challenging philosophical quests. In recent times, inspired by S. Chandrasekhar (Nobel laureate, 1983) on the life and death of stars, the physicists who had agreed on the beginning of the universe now had to work out its end, but they need to know the nature of dark matter and energy to find the weight of the universe; otherwise it will keep expanding into the last throes of its death. And, of course, the illustrious President Barack Obama had taught at the University of Chicago Law school in the 1990s. A Chinese may quibble: The University of Chicago must have the right 'fengshui' for inspiration and research, despite being in the heartland of the city's poor.

The year 1968 was a watershed year for the American nation, full of landmark events — the assassination of Martin Luther King, Jr. in April. King dared to have a great dream for America while we lesser beings remain imprisoned by our inconsequential dreams. Only two months later came the assassination of Senator Robert Kennedy. Across the nation, the citizens were sadly divided by the unnecessary Vietnam war, with many supporting the anti-war campaign of Democrat Eugene McCarthy. I could not help, but be engrossed in the freedom of thought and the great societal changes in the US national environment. However, I remember having to keep indoors in fear, with ears tuned to the radio, as riots broke out in downtown Chicago following King's assassination. This certainly could not fit into the American dream!

At the University of Malaya, where new qualified staff were required for the growing Department of Chemistry, our former lecturers and professor had always been on the lookout for their own scholars. In early 1968, positions for lecturers were advertised in local and UK papers, and both SH and I received invitations to apply. Some months later, we were offered positions as lecturers while still at Chicago. At that time most Malaysians would return to serve their own underdeveloped nation with never a second look at greener pastures elsewhere. Somehow with modest but free opportunities, the political harmony of those days and abundant sunshine made Malaya something of a tropical paradise, free from hazards like that of Chicago's racial strife or London's stressful hustle and bustle lifestyle and unpleasant weather. It had also been in my hope and dream to return home for a career. By that time SH had defended his thesis and the degree was at hand. We were dating

Beyond The Ph.D. and Chicago

seriously at that time, and with our parents' blessing, we decided to get married before we returned home. In our minds, it was to avoid the wedding 'complexities' that were bound to arise when families of the two parties live in different towns — for us it was Ipoh and Penang. It was only on return that I realised that a home wedding among family and close friends could have been easily arranged without too much inconvenience. Instead it was among a 'chemistry' family and fraternity that we had our wedding. We were married in the chapel of the University of Chicago in a simple Catholic ceremony, witnessed by my bridesmaid Joyce, my laboratory and hall mate in HKU, then migrated and settled in Toronto, and best man, David, also my Year 3 course mate in HKU but then SH's Ph.D. contemporary in UC. J. R. Johnson, postdoctoral in laboratory D18 of UCL, and his wife, Maryann, came from Kansas to attend. The two groups from Professors Halpern and Closs all graciously attended. We are much indebted to Mrs. Kharasch for hosting our wedding reception in her lovely three-storey apartment. She was the widow of the renowned free radical chemist M. S. Kharasch of the University of Chicago, with whom Rayson Huang did research. (see 'Wedding, Chicago, 1968', overleaf)

Thus, the return journey was a sort of honeymoon. We travelled *via* Europe; in those days of air-travel, multiple stopovers in the journey were permitted, so long as the total mileage did not exceed the mileage allowance between the two points of travel. Our first stop was London; for SH it was a first visit and for me a chance to visit my Alma Mater, especially to meet Tobe and Johnson again. Our next stop was Rome — and we arrived amidst an airport baggage-carriers' strike and SH had to haul our luggage from the hold of the plane! However, this little inconvenience was more than compensated for, as we enjoyed the treasures of Rome in a limited three days' stay, touring the Vatican Museums, the Sistine Chapel to see Michelangelo's

Wedding, Chicago, 1968

With Prof. & Mrs. J. Halpern

With Mrs. M. S. Kharasch and Joyce Lee

With Prof. & Mrs. G. L. Closs

ceiling, St. Peter's Basilica to see the famous Pieta, the Coliseum and The Forum for a glimpse of the Roman Empire. The Fall of the Roman Empire must be a great lesson for mankind — that a vast and mighty empire of more than a thousand years' duration could fall! A calamity which historians had ascribed to a combination of factors, such as the rule of despotic emperors, rise of Christianity, monetary troubles, military problems, decadence and a lack of interest in the sciences.

After that it was on to Hong Kong where we paid our respects to my relatives, serving traditional ceremonial wedding tea to my uncles and aunts, and to enjoy the culinary delights of HK. Finally reaching Kuala Lumpur, we were met by my loving parents who had come from Ipoh to house-hunt and rent on our behalf, a spacious double-storey semi-detached house in PJ, a satellite town of KL, close to UM.

CHAPTER

9

A Beginning in Academia

It was with a sense of hope and aspiration and a will to serve our young nation in its only institution of higher learning, that I assumed duty as a lecturer in the Department of Chemistry, UM, on 15 August 1968 (see picture overleaf). The university was founded on the model of British universities of that age. Hence the hierarchy of academic staff was Assistant Lecturer, Lecturer, Senior Lecturer, Reader (for a staff of professorial merit but with no available chair) and Professor. Traditionally each department had only one Professor who was also head of department (HOD) by virtue of his appointment. Professorships were appointment posts, not promotional positions. Rayson Huang was the founding Professor of Chemistry and was HOD since 1959. 1968 had seen the retirement of the founding Vice Chancellor Professor Alexander Oppenheim (renowned mathematician of Oppenheim Conjecture fame, 1929) in 1965, followed by two Acting Vice Chancellors, namely Rayson L. Huang and Chin Fung Kee, Professor of Engineering, each serving one year until Dr. James H. E. Griffiths, distinguished physicist from Oxford, was appointed in 1967. With Malayanisation, he was succeeded by Professor (later Royal Professor) Ungku A. Aziz, Professor of Eonomics, who served for two decades

UM Chemistry Building, 1960s

(1968–1988). During that period, interaction between him and Chemistry academics was scanty. Except for the HOD, the academic staff met him only in formal settings, e.g., at interviews for tenure and promotion. After him, UM had seven Vice-Chancellors over the next twenty years; the incumbent Professor Gauth Jasmon, appointed in 2008, is an engineer, who had previously held the position of the President of Multimedia University Malaysia (Malaysia's first private university, established in 1994).

In 1969 Rayson Huang left to take up the position of Vice-Chancellor of Nanyang University in Singapore and two years later of his Alma Mater, the University of Hong Kong. Francis Morsingh, holding a personal chair on Natural Products, became HOD for a brief two years, before he left to head the School of Chemistry in the new University of Science in Penang. Chan Kai Cheong was the next Professor of Chemistry and HOD. In the late 1970s, the staff hierarchy was 'Americanised' somewhat to Assistant Professor, Associate Professor and Professor. A promotion from Assistant to Associate professorship was considered

A Beginning in Academia

after seven to eight years of service in my time, but this period was reduced to five years in the 1980s. Promotion to the next level did not follow any time frame. Despite this new staff structure, the previous professorship-by-appointment position still remained, being dependent on vacancy made available by a resignation or retirement of an incumbent professor.

Since the early formative years, the University had in place a system of external examiners for its examinations. To establish the standards and set up the credentials of the new Department of Chemistry, UM on the recommendation of HOD Rayson Huang, had appointed for external examiners eminent professors of the stature of Lord (later Baron) Alexander R. Todd of Cambridge University (Nobel Laureate, 1957, for his work on nucleotides and nucleotide co-enzymes), H. J. Emelius (Inorganic Chemistry, Cambridge University) and Arthur J. Birch (of steroids and Birch reaction fame, FRS, founding dean of the RSC, ANU.) In later years, we had three external examiners on a concurrent term of three years, for each of Inorganic, Organic and Physical Chemistry; they visited us during annual examination periods on a rotation basis. That system had worked out very satisfactorily. Review of examination papers, together with reviews of syllabi during the visits of the 'Externals' kept our standards in check and ensured that they were on par with those in the major British Universities. Moreover, it was very beneficial for students to interact with them, and attend their research seminars, much like my personal experience in HKU. With the availability of British Council sponsorships in the late 1970s, I was able to recommend that my two previous supervisors from UCL (Martin Tobe and Michael Johnson) spend short-term visits in our department on two separate occasions. Following the visit, Martin Tobe served as department external examiner for a 'term' of three years, and was succeeded by A. G. Sykes of the University of Newcastle-upon-Tyne.

Unfortunately, since the 1990s, the system of external examiners was partially abandoned, and finally abolished and concurrent with this the coincidental decline of the University began.

In the work place, we became junior colleagues of our former lecturers/mentors. Upon arrival, I found on my office door, the name tag '*Goh Lai Yoong*', though I had applied for the position as *Wong*. At the time, I did not make any stand on this matter, though I had decided on 'Wong' when Professor Halpern had earlier asked me about my name preference in publications. As a result, I now have publications under both surnames. I believe it is more appropriate for professional women to retain their maiden names (what more with such a high percentage of divorces nowadays!). I was the second female lecturer in the department, after Su Eng Loke (née Huang), an Adelaide Ph.D. in Organic Chemistry. The academic staff then counted less than twenty and included nationals from UK and Australia. The early 1970s saw a large influx of new colleagues, with the new recruits coming from UK, Canada and Australia, in addition to the locals. By the 1980s the change in the medium of instruction from English to Malay and termination of expatriate allowances and privileges, rendered it unattractive for the overseas staff to stay, as the wages were relatively low, accentuated by the weakening Malaysian ringgit.

The academic year had started in May, and I had already been assigned courses to teach. My *maiden* duties consisted of the teaching of a Year 2 Practical Inorganic course and two courses to an Honours Year class of about fifteen students, namely a core course on 'Mechanisms of Inorganic Reactions' and an optional course in a topic of my choice, 'Insertion Reactions'. The Honours degree followed an elitist 3+1 year system; so the class cohort, initially in the teen numbers, seldom exceeded fifty over the years. I was mainly involved in the teaching of Inorganic or

Organometallic lecture courses at Years 3 and 4 (Honours) levels. Lee Soo Ying was in my second batch of Honours students, graduating with First Class Honours, literally without the need for any tutoring and was to be promptly selected by the University of Chicago for the Ph.D. programme.

Though still settling in, we had to find time during weekends to proceed to Ipoh and to Penang, to perform our wedding formalities — ceremonial tea-serving to senior relatives and belated wedding dinners for relatives and close friends. At the event in my home town Ipoh, I was glad that Professor Ho Peng Yoke first congratulated my parents before me on my Ph.D. degree. That was in accordance with the Chinese tradition of accreditation of merit — that we should realise that our success derives from the contribution and sacrifices of our elders. I was fortunate that mine did not possess the traditional business mentality, which failed to understand why their children needed to spend so many years for knowledge or dreaming about some area of knowledge, without expectation of much pecuniary returns. In years to come I faithfully taught that wealth of knowledge is mostly not measurable in monetary terms, but that it will bring forth a richness and fulfilment of life.

For a new lecturer, research infrastructure also had to be set up. It all had to be hands-on as there were no research students nor honours-project students. There was also no such thing as present-day 'start-up' grants, available in world-class universities. Small 'things' like glassware could be bought under departmental teaching funds. To help defray expenditure for some special research items, I had applied to the Royal Society of Chemistry London for small research grants of 100–150 pounds sterling; such grants given in 1972, 1975 and 1990 were much appreciated and put to good use. On the upside, the department had a very competent experienced glass-blower Rahman, who was of great help. He could make for me a rather complicated Schlenk line of the like of what I used in UCL. At this juncture, I must pay tribute here to the group of four dedicated glassblowers in UCL chemistry department. Our requests were completed very quickly — normally in just a couple of days — their joy was to see their 'creations' in use. UM chemistry department was fortunate to have one of them spend a couple of months around mid-1970s in our department's glass-blowing workshop to interact with Rahman and his assistant.

Research manpower support had always been minimal during my tenure in UM. Even as the academic staff establishment increased to above forty, the total number of tutors were less than twenty; these were the full-time graduate students, also serving as demonstrators in undergraduate teaching laboratories. Their assignment to a faculty member very much depended on the HOD, and because the number was so small there was not much choice and they were normally directed to a small select group of projects. In my twenty-seven years in service, I had barely benefitted from this small pool of tutor-researchers. The few research students of mine were external and part-time, like Karen Krouse Badri, an academic at Universiti Pertanian Malaysia, or supported as research assistants, like Tay Meng San and Richard C. S. Wong (partly), on my small grant, when manpower expenditure became permitted from mid-1980s. This meant that a lot of bench work and running of instruments had to be self-conducted, DIY from *a* to *z*. Amongst the ridiculous extremes were: buying the dry-ice for a cold reaction or trap — fortunately there happened to an ice-cream shop not too far from the university (the vendor must have wondered what this woman did with each block of dry-ice!), going personally to a metal foundry shop to get a vacuum pump support rack or other shelves made, and finally packaging and tying up the finished manuscripts and, occasionally taking them to the

A Beginning in Academia

post office, as well. I consider myself lucky to have had the help of a very competent laboratory assistant, R. Chiam, for a few years till he migrated to Australia in the early 1980s. In the department, these assistants were assigned to teaching laboratories and to a lecturer-in-charge, to assist in research outside of laboratory teaching hours.

Fortunately also, all the departmental facilities were available 'free' for usage. We did manage to acquire a 60 MHz nuclear magnetic resonance (NMR) instrument, supposedly for the teaching of Honours students and years later, after much tough negotiations with university authorities, a 100 MHz NMR instrument, a medium-resolution mass spectrometer and an electron-spin resonance instrument. It was a time when the teaching staff and laboratory assistants had to be involved with instrumental maintenance as there were no available positions for special technical staff. When obsolescence overtook some of these instruments in the late 1980s, there was a drought of new instruments, especially high-field NMR and FT-IR instruments because of cost and maintenance. However, some time in the early 1970s the department was generous enough to buy for principally my use a *stopped-flow* instrument for fast reaction kinetics, thanks to the support of Professor Chan (HOD). I did whatever I could with the assistance of Chiam. My first independent papers (1970–1972) were co-authored with my husband, arising from our common interest in free radical chemistry [mine, inorganic in nature on chromium(II) and cobalt(II), and SH's, organic in nature on bridgehead carbon radicals e.g. 1-adamantyl and dimethylbicyclo [2.2.2]octyl] (illustrated below). With no research hands

 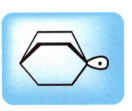

available, in this DIY situation, we realized we could not be too ambitious; still sometimes this realization did not alleviate the disheartening feeling to find one's very ideas worked out and published before one could even import the chemicals to start. We all realized that there are no second prizes in scientific discovery even though there was no publish-or-perish culture. Such incidents of research intentions dashed and dreams unrealised were to recur now and again, with only virtual reactions in the mind and on paper!

Outside the ivory tower, two memorable disastrous events occurred in Kuala Lumpur during these years, namely, the May 13 politically-influenced racial riots of 1969 and the once-a-lifetime great flood in 1971 after a week of incessant rains. Work at the university however was unaffected but it became evident in later years that the face and freedom of Malaysian universities were to be changed drastically from the implementation of new government policies.

CHAPTER

10

Sabbatical Leaves

In an 'isolated' environment away from the greater centres of research activity, it was fortunate that the University had kept the tradition of granting full-salaried sabbatical leaves to tenured staff for research, but unfortunately for me there were discriminatory policies in entitlement of allowance to women staff with husbands serving in the same institution. This gender discrimination in sabbatical leaves persisted to the time of my retirement. This was also practised in job remuneration, which could result in 30% difference in total take-home pay for the woman. However, this pecuniary shortfall faded into a non-issue, considering that it was my very own choice to remain as an academic in the country's only ivory tower at that time, and realising that an obsolete bureaucracy would take a long time to change. Indeed, equal-pay-for-equal-work was finally realised only in the late 1980s.

Arising from such policies, overseas sabbatical leaves had been expensive undertakings for me, especially when our children grew to primary-school age. Having them with us meant that we invariably brought along a maid (1981–82, in Canberra) or my mother-in-law (1986–87, Texas A & M), so that I could devote more of my time to research. Getting a VISA (temporary category) for an accompanying maid had required a lot of convincing at the Australian Embassy in 1981. Despite all these hurdles, e.g. financial, children's education and immigration, we had considered it necessary to go to a research-active centre in order to keep abreast with the forefront of research, considering that Kuala Lumpur was very isolated from global research centres and even locations of international conferences. For me any change from a usual monotony of lifestyle was much welcome as new experiences always did not fail to stimulate and rejuvenate. Likewise the children had benefited from the exposure to different educational styles and cultures.

To Chicago Again

We had invariably availed ourselves of the 3-years' rather than the 5-years' option of sabbatical leave of duration of five and nine months, respectively. For our first such leave, it was SH's choice to spend some time on cancer research in the laboratory of R. G. Harvey, associated with the Nobel laureate C. B. Huggins of the Ben May Laboratory of Cancer Research in Chicago. Thus there came an opportunity for me to spend a second term in Halpern's laboratory. In February, 1972, we returned to Chicago, flying *via* Tokyo to San Francisco, en-route to visit the Durrum HQ for stopped-flow instruments in Palo Alto. Strangely we were to learn from this part of the world that the University of Malaya had the distinction of conceiving Parkinson's Law,* the scourge of many government bureaucracies.

---

*C. Northcote Parkinson was Raffles Professor of History at the University of Malaya, when he published the article in *The Economist* in November 1955 — '*an observation that work expands so as to fill the time available for its completion*'.

Sabbatical Leaves

On this journey we travelled to Vancouver to visit a LHT hall-mate Poh Yu, (then librarian in Simon Fraser University) before proceeding to Chicago. I was much welcomed again and had the opportunity and joy of working in the new chemistry building. The new laboratories were spacious and extremely clean, one factor being the luxury of filtered incoming air. This reminded me of the need to wipe bench tops in the Searle laboratory clean of dark dust (arising from dirty coal used for heating mostly poorer city homes) every morning before I started work in 1967/68. In the design of this new building, the ease of access to functional rooms, e.g. NMR, stores, etc., was very well considered. My brief stay here overlapped somewhat with that of J. E. Bercaw, a talented chemist, who left to join the faculty in CALTECH, where he became an eminent professor. Here in Halpern's laboratory, I had the opportunity to share with him many laboratory techniques and end with a joint paper on cobalt(I) chemistry. We returned to Malaysia *via* London where we met up with Tobe and Johnson, and visited the new chemistry laboratories, recalling to mind that in D18 laboratory in which I carried out my Ph.D. project, stiletto heels could possibly pierce through some of the floor boards! In this city SH met 'Edward Arnold's representative', the publisher of his book, coauthored with R. L. Huang and A. S. H. Ong, *The Chemistry of Free Radicals* (1974, Edward Arnold London); next was to Germany to see the EXPO in Munich followed by a detour to Köln (Cologne), where SH visited Professor E. Vogel. We took the opportunity to stopover for two days in Venice to experience a gondola ride in one of its picturesque canals, and visit Murano Island, a short vaporetto (water taxi) ride from Venice. I particularly wanted to see this home of artistic glassblowers at work in their glass factories. Seeing them work and the lovely objects of their creation reminded me of the efficient and happy glassblowers in UCL Chemistry Department and the objects and figurines that adorned their benches in the workshop. Our next stopover was Athens, capital of Greece, one of the world's oldest cities, spanning 3400 years, noted for its philosophers, Socrates, Plato, Aristotle and others. In our brief visit to this city of antiquities, we saw just a little of the numerous monuments on the Acropolis, which is the greatest and finest sanctuary of Ancient Greece, built on a hill, The Sacred Rock of Athens. Yet again in this part of the world, it was difficult to imagine that a civilization with rich culture, education and knowledge of early science could succumb due to human failings.

To Cornell University

For our second sabbatical in 1976, as SH was interested in working in J. Meinwald's laboratory in Cornell University on insect secretions and pheromones, I searched for someone of suitable interest to work with in the same university. In those pre-*www* days, the one comprehensive source book for this was *The Directory of University Research in USA*. I located Earl L. Muetterties (ELM), who offered me a laboratory bench with a small stipend and a temporary research position. But it was a tough decision to make regarding placement of our two very young children — son Chih Chien (CC), barely three years old and daughter Sue Yen (SY) one year old. We had been managing with the help of a full-time live-in maid, with my mother-in-law overseeing during times of changeover of maids. My research activities had naturally slowed. When SY was born, I received a congratulatory message from Michael Johnson with the comment that 'a baby is equivalent to 3 research papers', what a magical number! Perhaps based on my personal productivity record so far? After much thought and arrangement, we decided to place CC in the charge of my parents in Ipoh with the attendant maid, Mui, a young capable girl of 20 years, and SY was to be with my aunt,

Sabbatical Leaves

Yan Sin, also in Ipoh. SH and I looked like going off on a second honeymoon, but with heavy hearts even though for only six months.

We departed for Cornell University in September of 1976. I felt a sense of unease on the flight from New York to Ithaca in a ten-seater plane. In this small campus town, apartments were scarce. We found only a studio apartment, within walking distance of the Baker chemistry building through some playing fields. There were but a few streets in this small town of less than twenty thousand, comprising mainly of university students. Cornell University is a large institution, offering undergraduate and graduate fields of study in nearly a hundred academic departments. Founded in 1865, Cornell is both a private endowed university and a federal land-grant institution of New York State. A member of the Ivy League, it is an institution with many firsts. It has been a leader in women's and minority rights since 1870. Notably, it has forty Nobel laureates affiliated with it as faculty members or alumni. Of these prizes, eight were for Chemistry (the earliest in 1936 for P. J. W. Debye and the latest for R. Ernst in 1991). We found that there was a large group of Malaysian students there at that time studying Agriculture, in which Cornell University excelled. We also met there a Biochemistry colleague, Dr. Balakrishan, also on sabbatical from UM.

Muetterties had moved from industry (DuPont) to academia in Cornell University in the early 1970s. Working at the interface of organometallics and catalysis, he maintained a productive group, then consisting of eight members, mainly postdoctoral associates. In his laboratory I was introduced to the stringent demands of hardcore organometallics, requiring absolutely dry and oxygen-free atmosphere and conditions. My previous work involving reactive chromous cations and pentacyanocolbaltate(II) anions in aqueous media under an inert atmosphere of nitrogen was comparatively less challenging. When I arrived in Cornell in 1976, it was two years after Cotton's paper had been published in the *Journal of the American Chemical Society* on the structure of the dinuclear molecule $[CpCr(CO)_3]_2$, which revealed an unusually long metal–metal bond. After a discussion on Cotton's finding, ELM and I decided to explore the effect of carbonyl substitution with a bulkier ligand, for instance, trimethylphosphite for a start. A post-doctoral fellow commented: 'A Cr system! That is a difficult system!' As I found out, such a perception of these systems actually arose from their lability, which could be capitalised upon, so long as complications could be obviated by prompt workups to isolate the primary product(s). Indeed, subsequent reactivity often led to elegant families of derivative complexes. Here I had my first taste of a labile fruitful system, organometallic and radical in nature. This brought to mind Michael Johnson's comments on the wealth of diverse reactivities of radical systems. In Muetterties' laboratory, I was very lucky, being able to work up the whole reaction system within six months for publication in *Inorganic Chemistry*, a journal of The American Chemical Society. We were both thrilled with the isolation of a new dichromium species, which he called a 'M–M bond/no-bond' species. It was in this laboratory in Cornell where I dreamt up more reactions with reactive systems, based on this exemplary species. Although ELM was himself away quite often in Berkeley in sunny California on sabbatical leave himself during my stay in Cornell, he had tremendous energy and there was still plenty of room for professional interaction, though some of the regular weekly group meetings (one lunch time and another after-dinner meeting which sometimes went well into the night) had to be cancelled. I had gathered that ELM was actually considering accepting an offer from Berkeley, and indeed, he left Cornell in the following year, but sadly he succumbed to cancer a few years later.

We spent a fall and a winter in Cornell, September 1976 — March 1977. During the winter months, the fields were beautifully white with snow, but it was messy on the streets downtown. This 'student' town practically closed during the Christmas vacation. It was understandable, because like the predominantly student community, we also left town with Dr. Balakrishnan. We travelled in his car to Toronto to visit my friend and former bridesmaid Joyce and her husband Yu Fai, and to see Niagara Falls on the way. It was a dramatic change to see the white scenery in winter after having seen the Falls in the summer of 1968. Much refreshed from gazing at the mighty waterfalls we returned to Cornell to continue research, at the end of which we returned home *via* Los Angeles, and were hosted for two days by Professor Lawrence Singer (USC) and family. Larry was SH's lecturer for a graduate course in Chicago. Our itinerary (the cheapest as proposed by our travel agent) took us to Noumea, capital of New Caledonia in the Pacific, enroute to Brisbane. We had not expected to find such a delightful tourist resort in this remote Pacific island, and so had only scheduled a day's stopover there. It was only after arrival that we learnt that New Caledonia is home to one of the largest lagoons in the world, ideal for water sports, like wind surfing, snorkeling and scuba-diving. From there we proceeded to Brisbane to visit my sister Lai Kam (a pharmacist) and family, who had settled there since 1972. That was our first visit to Australia and we were happy that my sister had settled in a very nice suburb of Brisbane, a lovely city with a lot of waterfront, since it is surrounded on three sides by the Brisbane River. With such a beautiful and safe environment it was opportune to explore this island continent for prominent places for future research or study visits.

To the Australian National University

During the 1980s, another sabbatical entitlement came along, which in our opinion should not be missed, although we were already established researchers in our own fields. This time, in addition to selection of the host institution on the basis of academic reputation, we also needed to consider the increasing needs of the growing children. We arranged to go to the School of Chemistry (RSC), Australian National University (ANU) (1981–1982), known to be one of the top institutions with very strong faculty, and Canberra being a well-planned medium-sized city would be ideal for the children. Indeed, we found both the academic and living environment excellent in all respects, and we spent another sabbatical there in 1990–1991. In 1990, I had actually been offered a Commonwealth Developing Countries Fellowship of the Royal Society for research in Professor J. Lewis's laboratory in Cambridge, but had to forgo that because of the start of our children's secondary schooling in Australia at that time. At the RSC, Swee Hock had worked in the group of A. E. L. Beckwith, while I had worked in the laboratory of M. A. Bennett on both sabbatical occasions, and also during four other short visits in the 1990s; I heard some at the RSC had called me *the permanent visitor*! Doubtless, I had benefited very much from my various visits there, e.g. from its excellent infrastructure and facilities, especially NMR spectroscopic facilities, its regular system of group meetings and seminars given by School members and the numerous visiting prominent scientists. A system of annual safety checks and enforced 'spring-clean' of all laboratories by the researchers, plus a very efficient daily cleaning service, ensured a level of orderliness and cleanliness unmatched in other chemistry departments in which I had worked, rivalled only by the new Chicago laboratories where I had worked in 1972. In the RSC, I literally was able to walk in and begin bench work the very next day. It was during my stay of 1981–82 at the RSC that I first met Ph.D. student Leung Pak Hing, present HOD in CBC, in the research group of Bruce Wild. There were common morning and afternoon tea sessions for the entire school; my observation was that these facilitated a very good social and cohesive spirit among the members, in addition to a fruitful exchange of news and ideas. Indeed, even as visitor, I had managed to cultivate a very good rapport with most of the RSC staff; I remember especially the friendly professors who included M. A. Bennett, D. P. Craig (who had chaired my Ph.D. symposium at UCL), Bruce Wild, G. B. Robertson, A. E. L. Beckwith, the late A. M. Sargeson and the late R. Rickards; the research fellows whom I had met, have subsequently all progressed in their careers: N. K. Roberts (UNSW), T. W. Hambley (presently Director of Research, Faculty of Science, Sydney University) and L. M. Rendina (Sydney University). The late Rena Chao, a member of the RSC since its founding and was head of Microanalytical Services, became a good friend and had hosted me in her comfortable spacious home during some of my short visits to the RSC. Likewise, a lasting friendship developed with Tin Culnane in the NMR technical group.

To Texas A & M University

Texas A & M University (TAMU) in small town College Station was another centre for a 7-months' sabbatical sojourn in 1986–87. Situated in Central Texas in the heart of a rich cattle and cotton-producing valley, the town owes both its name and existence to the university's location along the railroad. With Bryan, it makes up the Bryan-College Station metropolitan area, which had a population about a hundred thousand in the mid-1980s. The city is located within the most populated region of Texas, near Houston, Dallas and San Antonio, three of the ten largest American cities. TAMU is one of the premier universities in Texas, and the flagship institution of The Texas A&M University System. It is reported that College Station was named by '*Money Magazine*' in 2006 as the most educated city in Texas and one of the most educated nationwide, on the basis of the size of TAMU. It is also a vibrant research centre, well-funded by national agencies and the Welsh Foundation, which had attracted F. A. Cotton to set up his 'home laboratory' in the Chemistry Department, which at that time also saw the relocation of Sir Derek Barton (1969 Nobel laureate) from Giff Sur-Yvette (France), where he was director of the Institut de Chimie des Substances Naturelles.

It was a long haul to College Station for the group of six of us, which included three children and my mother-in-law. We flew via Los Angeles to give the children an opportunity to visit Disney Land. Next was a time-consuming journey southwards and finally we were met by a group of members of the Malaysian Students Association of TAMU, to drive us on the last leg to College Station. Settling down activities included looking for an apartment with the kind assistance of Maria Ludvig, a Ph.D. student in the group of Marcetta Y. Darensbourg. After signing the rental agreement and getting all facilities fixed up, we moved in after two days. There was no time to be wasted in organising the children's education, and we promptly registered them into the local schools within walking distance of the apartment. In this small town environment we were happy that they could benefit much from their two terms in the school, which even catered to the continuation of their music education (piano, viola and violin) during that short period, and also gave daughter SY an opportunity to participate in the inter-schools Mathematics competition, in which she had excelled. In the occasional social gatherings with the Malaysian students, my mother-in-law would surprise the Malay students with her perfect pitch of Bahasa Melayu, which she had acquired from harmonious relationships (during pre-independence and early post-independence days) with nearby 'kampung' (village) folks in Penang.

In the Chemistry Department, SH worked with Martin Newcomb, while I was actively involved in Marcetta's group, enjoying work in the laboratory, as well as visiting her homestead where she and Donald kept horses. I had a good rapport with the research group. Wen F. Liaw from Taiwan, a first year Ph.D. student was very helpful and drove us to Austin for our return flight. He is now an established influential scientist in National Tsing Hua University, Taiwan, and I was very happy to have him speak at '*my commemorative*' symposium in 2007. The Chemistry Department of TAMU would have been considered isolated in the deep south of US, if not for the very regular visits of eminent chemists invited to give series of lectures. Seminars were also scheduled for short-term visitors like myself. During that occasion I became acquainted with my seminar chairman Rinaldo Poli (now in the University of Toulouse), research associate of F. A. Cotton; I was to meet him again at the 33rd ICCC in Florence in 1998, and again at the Mark Vol'pin Memorial International Symposium in Moscow in 2003.

On a weekend (refresher) trip to Austin with Wen, we learnt much of the history of Texas. We spent a Christmas holiday in the historic city San Antonio, which had impressed us with pleasant sights and a nice environment, in agreement with what many had described as Texas's most beautiful city. In March 1987, after seven months in College Station, it was time to travel home *via* Taipei, which struck us as a bustling and lively city but a little dusty and dirty. Our three days' stopover gave us all a brief glimpse of life and culture in Taiwan in the early years of its economic transformation.

CHAPTER 11

Research Reality in The Home Environment

With early research training in investigations on very basic processes in inorganic/organometallic chemistry, viz. the formation and cleavage of carbon–cobalt bonds in organocobalt complexes, I had inculcated more interest in basic research than in the applied. As an inorganic chemist, I see vast opportunities in the chemistry of the non-carbon elements in the Periodic Table. As an organometallic chemist in the 1960s, when Vitamin B12 coenzyme was the only stable σ-bonded cobalt–carbon complex known and a naturally-occurring one for that, and when chemists were attempting to find the rationale for the instability of metal-carbon bonds, I saw in my mind's eye long and varied avenues of research opportunities for the chemistry of compounds of carbon with its metal or non-metal counterparts in the Periodic Table. Thus it was that my earliest attempts at independent research revolved around the reactivity of pentacyanocobaltate(II) with bridgehead carbon species and of various chromium(II) species with arylalkyl halides. These cobalt(II) and chromium(II) species behaved like radicals towards organic halides, forming organometallic compounds.

As I had mentioned above, my short research stint in ELM's laboratory led to an encounter with a radical species of another kind — an organochromium 17-electron species derived from dissociation of a dichromium molecule. An observation of its high reactivity gave rise to new ideas of possible research pursuits. But that would require a Drybox, costing to the tune of sixty thousand ringgit, a princely sum at that time. That was not easy to obtain in a grant-poor institution. The time had not arrived for individual researchers to apply for grants of the magnitude of thousands of Malaysian ringgits. A friendly colleague commented that I was asking for the sky — better that I be contented to do (in his view) cheap 'simple' chemistry! Actually, for some less demanding systems, I was already pushing experimental skills to the limit with endless argon/nitrogen lines, syringes and Schlenk glass vessels. Another more senior concerned colleague advised that it would be better that I concentrated on looking after the children. Perhaps *I was dreaming,* as I recalled the wisdom of my secondary school teachers at CHIJ Ipoh, who had advised 'to hitch your wagon at the stars', for then it will fall on the roof top, and *I later envisaged* the consequence of 'hitch it at the ceiling and have it fall crashing on the floor'. Actually I was only being enthusiastic rather than being ambitious, for without supporting hands other than my own pair, it was likely that all things would just be like in a dream, or at best I would only be pursuing a hobby! Fortunately, Professor K. C. Chan (HOD) must have believed in me, because two applications later, I had approval for the purchase of a Vacuum Atmospheres Drybox, which was commissioned in 1980. It was only from the mid-1980s that funding for research was regularised under the Malaysian Ministry of Science and Technology; funds then became available for instruments and consumables, but not manpower until much later. For some bureaucratic reasons, grants also could not cover maintenance of departmental instruments, which must be applied for on an *ad hoc* basis, as and when an instrument broke down! Even then requisitions for repairs were by no means automatically granted. If the department could not afford to cover the expenditure, an application had to be made to the deputy vice-chancellor of research, but considerations for approval would have to be prioritised, based on the number of users of the instrument, be it an NMR or MS instrument, or any other. All researchers suffered silently when there were disruptions to the progress of any project but some could resort to innovative or unconventional ways to carry on.

For me the period from mid-1970s was a taxing time. Near the end of that decade, a third child, daughter Sue Lynn (SL) joined the family. Indeed, attention and activity were required on several fronts — teaching, research and family. I had to see to the needs of three young children but was fortunate to get the assistance of the same efficient maid Mui, who stayed seven years with me till she married. My sister Lai Fong, graduate chemistry teacher at La Salle Secondary provided invaluable support, as she provided transport for my children together with her own to and from school (the boys in La Salle and the girls in the neighbouring Assunta Convent School). That left me only the job of ferrying the children to extracurricular activities in the afternoons; thus I often had to leave office to do some chauffeuring and then return to do reaction workups. Such was the advantage/privilege of academic freedom wherein work hours were slightly flexible. In this respect, my newly-acquired Drybox facility with an incorporated freezer was a godsend, as it ensured that my primary reaction products did not suffer any irreversible transformation, till I found time to work them up for full characterisation. My students M. S. Tay and R. Wong also found that the Drybox was a boon for their work on the highly reactive cyclopentadienyl chromium tricarbonyl dimeric system. Its tremendous proclivity to cleave nonmetal-nonmetal bonds in the polynuclear molecules of non-metals of Groups 15 and 16, unfailingly yielded new compounds, many possessing novel structural features. Of special interest was a unique assembly of a core of ten phosphorus atoms, bonded to five chromium groups (note that elemental white phosphorus, naturally occurring in tetrahedral P_4 aggregate form, is spontaneously combustible in air, and had lately been trapped in a self-assembled iron cage for release on demand (see J. R. Nitschke, *et al.* in *Science*, 2009, 324, 1697–1699: *White Phosphorus Is Air-Stable Within a Self-Assembled Tetrahedral Capsule*). Equally unusual was the formation of a Cr_2P_5 triple-decker metallocene, as well as Cr_4 assemblies formed by interaction (rearrangement without fragmentation) of P_4S_3 and P_4Se_3 cages with the chromium radical. These various reactivity features had been presented in conferences, and detailed in publications. The molecular structures illustrated below are meant to give the general reader a glimpse of the structural diversity of newly created compounds in this chemistry.

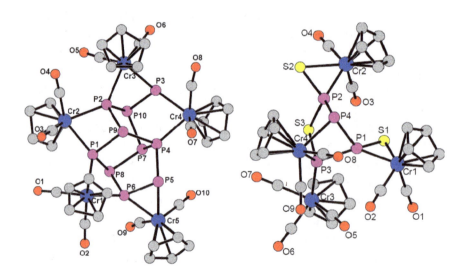

Research Reality in The Home Environment

The preparation of manuscripts was certainly not an undertaking to be attempted during the day, truncated intermittently by lecturing, tutoring, laboratory classes, student queries and family obligations. This had to wait till nighttime, when the household chores were completed and the children's needs attended to and this activity would often run from 10 pm onwards, with no 'Cinderella' running-off at the strike of twelve.

Working in an environment with minimal research facilities and technical support was really an uphill drive all the way. This was a time and place where research seemed non-priority, though unintentionally, as support was lacking from all quarters. On the library front, new journal volumes took 2–3 months to be available to readers. In those pre-*www* days, without any internet and all its attendant easy-access conveniences, peering through the fine print of the then weekly 'Current Contents' was a necessity; thus we would make note of new publications and send out request cards for those papers. On one occasion an eminent London professor replied *'in these days of easy photocopying, we do not send copies of our papers anymore'*. Really, who could have known that in Malaysia, though already above 3rd world status, there could be no original printed article to photocopy from? So even to these days in the age of the internet, I have always obliged on such requests, though rare now, by just sending a PDF file. Rightly, a professor at Texas A & M (mid-1980s) commented to my husband: 'I marvel at the way how you people manage in doing research, with no easy access to *all* literature, *outside* the circle of manuscripts-reviewing, which certainly told a lot of current trends, etc'. He was only referring to the intellectual and vital peer interactions. Yes, we were certainly disadvantaged in those aspects and others as well, compounded by the lack of foresight and poor intuition of leaders in developing countries, and a culture which did not value human resources and talent development at that point of time. That was more than a decade before the onset of ranking of universities and departments. Thus, it was a pleasant surprise for me and my colleagues to read in a 1975 edition of the *Chemistry & Engineering News* (American Chemical Society) an unsolicited 'rating report' of our department in UM as the best in the region, 'considering the quality of faculty, research facilities and research output.' That American professor must have been cognizant of our publications in leading journals like *Nature* (J. B. Lowry 1973), *Journal of the Chemical Society, Chemical Communications* (K. H. Lee 1970, S. H. Goh 1972, Ng Soon 1972 and 1974), *JCS Dalton, Faraday and Perkin Transactions* (numerous publications), *Tetrahedron* (H. H. Lee, K. H. Lee) and *Journal of Solution Chemistry* (K. H. Khoo), and book *The Chemistry of Free Radicals* (R. L. Huang, S. H. Goh and S. H. A. Ong. 1974, Edward Arnold, London). As American professors seldom came to this part of Asia in that period, we surmised that he was a private guest of one of two American chemistry professors (teaching at that time in the MARA Polytechnic, now a University of Technology, one of them being Gilbert P. Haight from the University of Illinois at Urbana-Champaign, and the other a Professor Wain), who frequently attended our departmental seminars. For UM and for the nation, it was an opportunity lost, when there were no subsequent concerted efforts to uphold that status. Could that have been sustained for the following 30 years to this day? I believe 'most probably', if as in other upper-rung global universities, more had been done to keep improving research infrastructure, recruit quality academics and provide support to research-active faculty in the form of, for instance, funding, technical support personnel, and prompt approval for maintenance and repair of research instruments, book/journal acquisition requests, and participation in conferences. In fact, the scarcity of conference funds limited the participation at international conferences for a faculty member to once in three years, with those publishing at the forefront being put in a lower priority list, because they were considered already well-exposed or established. That was indeed a trying research environment. Present faculty are fortunate that such a dearth of research funds no longer exists, as the universities today conscientiously strive to climb in global ranking.

Working in the confines of the Third World also meant that good research results came under incredulous scrutiny, especially in a competitive area and could evoke crude uninvited responses in referee reviews. But the worst responses could come from foreign competitor-colleagues. For instance, in the late 1980s, I had the unpleasant unforgettable occasion to receive an unwelcome distasteful response from a member of the international community — a prominent professor — to one of my most recently published papers in *Organometallics*, a major journal of the American Chemical Society. I am sure he would not have written his hasty condemning letter, had he noted the relevant dates of publication — namely of submission/acceptance, of a preceding paper, if any, whether in Communications or Conference Abstracts. In reply to my lengthy response to him, he implied that I had misunderstood him and kindly offered me space and facilities to work in his laboratory, something my peers interpreted as the equivalent of a gracious form of apology. He had by then realised that mine was but a *one-woman* team working in a seemingly obscure university. Unfortunately for him, I had already circulated *my response to him* to some of my Chemistry friends and colleagues in the international community, a step that I had considered necessary, on account of his threat to 'warn the Editorial Boards of...'. Subsequently, I had been asked on a couple of occasions for a read of that reply of mine to him, which I had not considered appropriate to acquiesce. Although the initial reaction to such unfriendly competition might be depressing because of unlevel playing fields, it did serve to bolster greater determination to continue the dream. Two years later, we met in person at a conference and he reiterated that *'we had a misunderstanding'*, perhaps gracious for someone of his stature, as I stated earlier. Subsequently, our paths had crossed on congenial terms on a few occasions when I was in NUS, and he had lately suggested that I should make a lecture tour upon my retirement. As I see it now, then was the age of Western dominance in the acquisition of new knowledge; but should there be a need to ask if any new science could come out of developing countries, much less Third World countries? Presently however, in the dawn of the Pacific century, I am sure that such recriminatory behaviours are not likely, at least less likely, to occur.

That unpleasant incident aside, I have had very encouraging interaction and support from many professorial colleagues in the international and local community. These helpful fellow chemists include: M. A. Bennett (RSC, ANU) and G. B. Deacon (Monash U.) for inspiring discussions; the X-ray crystallography experts — T. C. W. Mak and Z.-Y. Zhou (The Chinese U. of Hong Kong), E. Sinn (then of U. Virginia, presently of Western Michigan University), T. W. Hambley (U. Sydney), who collaborated with me until UM Chemistry Department acquired its own X-ray diffractometer, when my colleague Chen Wei could take over, and lately A. H. White (U. Western Australia); P. A. W. Dean (U. Western Ontario) when there arose a need for a selenium-77 NMR; K. Murray (Monash U.) who had always obliged with magnetochemistry problems; R. D. Webster (then of RSC, ANU, presently of NTU) for ESR and electrochemical investigations; V. I. Sokolov (Russian Academy of Science, Moscow) on a new area of spin-adduct Chemistry of C_{60}; and R. T. Boeré (U. Lethbridge, Canada) on the relatively unexplored organometallic chemistry of inorganic heterocyclic radicals. Departmental colleagues who had co-published with me included Y. Y. Lim and C. K. Chu.

As I stated earlier, chair positions at the University of Malaya were much restricted during those days. It was only after Chan Kai Cheong retired in 1986, that the Chair of Organic Chemistry fell vacant, whereupon Swee Hock applied and was appointed its next Professor. With so many 'hitches' and transformations in leadership in a system rocked by turbulent years, which saw three Vice-Chancellors between 1988 and 1995, I did not get promoted to professorship status till 1992, which for me was already close to normal mandatory retirement.

CHAPTER 12

Overseas Conference Participation and a Bit of Travel

Participation at conferences, particularly at international levels, is today considered a crucial part of frontier research activity. Nowadays, even graduate students in reputable universities are supported to participate in poster presentation. Unfortunately many obstacles had to be faced during my tenure in UM; despite being sited in a research-isolated location, support for such activities had been very low on the agenda, as mentioned earlier. I have always maintained the firm belief that interaction with overseas peers at conferences is important. As research progressed to a productive stage, I had availed myself of as many opportunities as possible to present the results in oral or poster presentations at conferences, despite being out-of-pocket as a result, being only partially funded by foreign and out-of-university sources, e.g. IKM and Lee Foundation. Thus, I had kept a regular attendance since 1984, when I was invited to participate in a memorial session for the late Professor Muetterties, who had passed away the year before. That occasion was a one-day symposium within the 188th ACS National Meeting in Philadelphia, and I was glad to meet again some of Muetterties group, especially Mrs. Muetterties and for the first time Patricia Watson (Muetterties' first Ph.D. student in Cornell University) and Victor Day (the collaborating crystallographer). On the home journey, I stopped over in Chicago to visit Professor and Mrs. Halpern, who hosted me in their lovely Hyde Park home, where I had previous occasions of enjoying their lovely annual dinner parties for the research group. I was very happy also to meet again with Professor N. C. Yang at UC.

The next few years saw regular invitations to conferences or symposia, where I presented papers on numerous aspects of my work on the reactions of the cyclopentadienylchromiumtricarbonyl dimer with non-metals in Groups 15 and 16 of the Periodic Table. In particular, I was almost a regular at the biannual International Coordination Chemistry Conferences (ICCC), e.g. 25th ICCC (Nanjing, 1987), 27th ICCC (Queensland, 1989), 29th ICCC (Gera, 1990). It was a privilege to provide the only beacon, albeit a small one, for Malaysia at such prestigious conferences. Being the sole representative of my country, I had been invited to the Planning Committee dinner functions. The main item on the agenda of such meetings was discussion of bids for holding of future conferences, as far as a decade into the future. Thus, I had the opportunity to be acquainted with the permanent secretary of the ICCC, Professor I. Reedjik, and it was good to help forge links for Malaysia and Singapore. In Nanjing, it was a pleasure to meet Martin and Rosalie Tobe, and that was the last time I saw Martin. That was also my first visit to China, and I consider myself very lucky to be able to visit one of the most delightful and historical of Chinese cities, with history spanning the time from the Three Kingdom's Period (220–280 B.C.), through the Ming Dynasty (1368–1644) to the twentieth century of the last emperor. I had looked forward to visit this capital city of several dynasties, this cultural city of memorials,

Overseas Conference Participation and a Bit of Travel

International Colleagues

B. Bosnich, Gold Coast, Queensland, 1989

J. E. Bercaw, Gold Coast, Queensland, 1989

Ekk Sinn and daughter April, in Kingston-on-Hull, 1997

V. I. Sokolov, Moscow, 2003

M. A. Bennett at SICC-2, Singapore, 2004

W. F. Liaw, Commemorative Symposium, Singapore, 2006

museums and historical sights, this city wherein occurred the Nanjing Massacre, as told by my mother. In 1989, travel to China was still forbidden by the Malaysian government. Hence I had to seek approval from our Home Ministry, which allowed me three post-conference days to visit Beijing. I flew in a local plane from Nanjing to Beijing, where as a lone tourist, I found that the little *'putonghua'* in my vocabulary was good enough to get me around the places of interest. My tight schedule of three packed-full days took me to see the Ming Tombs in Changping County, en-route to the Badaling Great Wall, about 30 miles from Beijing city. This is a section of the Beijing Great Wall, built around mid-sixteenth century during the Ming Dynasty (1368–1644). I was told that this is the most visited and best preserved section (about 400 miles) of the Great Wall of China, a zigzagging structure of amazing proportions (5500 miles). I had only half an hour to walk to see a few of the city wall platforms and countless towers. I left marveling at the capabilities, endurance and sacrifice of those ancient engineers and workmen, in the construction of that magnificent and monstrous edifice in such a harsh mountainous terrain. Such perseverance (forced or voluntary) of the men in those forgotten days somewhat eclipsed most present day creations, made possible by the help of technology and all forms of instruments and machines. Then I recalled the rude awakening on the previous day by the sight of a man pulling on his shoulder a cart holding a huge timber log (about 8 feet long by almost 2 feet diameter). That was during a side trip from Nanjing to Yangzhou, showing the down-to-earth state of affairs in China in late 1980s. What a difference two decades of determined efforts make to the progress of a nation!

My journey to the 29th ICCC (Gera, 1990) turned out to be most eventful. Going to conferences, I had always travelled alone. I had flown from KL to Frankfurt with a stopover in Paris. The forward journey from Frankfurt to

Overseas Conference Participation and a Bit of Travel

Jena by train was unintentional, after being mis-directed to the *ICC* venue in the city when I enquired for *ICCC* upon arrival at Frankfurt airport. A language problem? Unfortunately for me that lady at the enquiry desk could not sense that an Asian woman like me was hardly likely to be a fan of cricket! Or maybe perhaps she was so surprised that this unlikely Asian lady was actually a cricket fan or a member of the *ICC*.

I realised the mistake when I saw *ICC* banners all over town, but there was not sufficient time to return to the airport to catch the designated bus to Jena. I was totally unprepared for such a turn of events and hence had not done any homework on public transportation, nor brushed up my half-forgotten German; fortunately I met an English-speaking Northern European man who showed me the way to the railway station. I thus had an opportunity to travel through the countryside in that part of East Germany and witness the poor underdeveloped environment. At that time, the Berlin Wall had just fallen. On the hours-long journey, I understood from a kindly fellow traveller in our bilingual German/English exchange that I had to change train mid-way, otherwise I would find myself back in Frankfurt! Imagine how relieved I was to finally reach Jena train station and saw signs leading to the 29th ICCC reception desk! In my scary little detour, I paid a mental tribute to those travellers or adventurers of olden and present days for their bravery in venturing across treacherous seas and burning deserts to unknown lands and of unknown tongues. I was welcomed by Professor Uhlig, chair of the conference. He was present at my talk in Nanjing (1997) and thus invited me for this occasion to Jena. At the conference, mingling among the many German participants, I had observed their excitement and anxiety, as they pondered on the reunification of their country in the wake of the fall of the Berlin Wall. I gathered that because of this, many would-be participants from USSR could not attend, because of a currency/financial problem. During the next few days there, I also learnt that Jena was hometown of E. Schrödinger and Karl Marx, and famous for its Carl Zeiss optics and the Schott glass factories. It remained for me to marvel at the greatness of German science and technology.

The purpose of my stop-over in Paris was to visit Sr. Anna, my French teacher and music/singing instructor at CHIJ Ipoh during my secondary school days. I was hosted for two days in her Convent in the heart of Paris, close to Rue du Bac, where she brought me to visit la Chapelle de la Médaille Miraculeuse, in which is displayed in a glass enclosure below an altar the incorrupt body of St. Catherine Labouré (the body was exhumed in 1933, 57 years after her death; the fine preservation of her body, as evident in her skin, has to be seen to be believed). In this church was also found the shrine of St. Vincent de Paul, the Apostle of Charity; under his patronage the poor, the needy and the suffering in Malaysia and Singapore are benefiting today via The Society of St. Vincent de Paul, which comprises of close to a million members spread among almost fifty thousand confraternities in more than one hundred countries worldwide. With Sr. Anna, I also spent a day at the Louvre Museum; we had ourselves photographed with the spectacular Pei glass pyramid entrance in the background. I understood that this is one of the largest and most visited museums in the world, housing about 35,000 works of art, among them was the world-famous *Mona Lisa* by Leonardo da Vinci. In one day, I could barely feast my eyes on even a miniscule section of the vast and varied collections. During this visit, I had to forego seeing most of the other attractions of Paris, which included the Eiffel Tower, Notre Dame Cathedral, Arc de Triomphe, Centre Pompidou and Champs-Elysées, places I had visited before in the 1960s while a student in London. On the return journey, I detoured by train to Nice on the Mediterranean

Overseas Conference Participation and a Bit of Travel

to visit Reverend Mother St. Paul, Principal of CHIJ Ipoh during my secondary school years, residing then in Maur House for retired nuns. Thus, I obtained a brief superficial experience of the lives of the nuns, who had ventured across the continents and oceans to teach in Malayan schools since a century ago. They are to remain a special people living sacrificial lives of an unusual era, albeit one coming to an end, being replaced mostly by generations hurrying to accumulate all things and 'enjoy' life to the fullest.

In addition to establishing ties with Australian colleagues and keeping up on the research front, attending conferences in Australia in the 1990s gave me an opportunity to present more work on chromium-phosphorus/arsenic chemistry, and chromium-organic thiolate chemistry, as well as allowing me to make side trips to visit my children studying in school or universities in Melbourne. Such occasions included the 9th RACI National Convention in Monash in 1992 and OZOM 93 in Armidale in NSW. In the Inorganic Symposium of the 11th RACI Convention in Canberra in 2000 to honour Martin Bennett on the occasion of his retirement, I presented a paper on ruthenium-trithiamacrocyclic chemistry, an area of research which was begun in his laboratory in the RSC. Other participations took me to The Second World Chemistry Congress in Brisbane in 2001, and the IC-03 Conference in Melbourne in 2003. Likewise my two visits to Monash in 1993–1994 under a Bilateral Science and Technology Collaboration programme at Monash University, were also very timely for me to catch up with my children. In this bilateral programme, I worked with Professors Glen Deacon and William Jackson, but it also enabled me to develop a very good rapport with other Australian academics in Monash, notably Leone Spiccia and Keith Murray, and including brief encounters with Professors Alan Bond and Bruce West, who was then close to retirement.

Next I ventured to Banff (Canada) in 1994, for the 7th International Symposium on Inorganic Ring Systems. I had been invited because of my recent publications on novel metal-phosphorus-sulphur/selenium ring systems arising from the reactions of phosphorus, tetraphosphorus trisulphide and tetraphosphorus triselenide with an organochromium species. It was very nice to have a Canadian member of the audience tell me that he was one of the 'historical' authors whom I mentioned in my presentation! On this trip, I had travelled *via* Vancouver to see my librarian friend Poh Yu and then to Calgary, where I met up with SH's friend, Professor Chang, a Mathematics faculty at the University of Calgary. He showed me the interesting sites of that vibrant city, including some of the sports venues of the 1988 Winter Olympics. I learnt that Calgary is Canada's third largest city, situated at the foot of the awe-inspiring Canadian Rockies, and close to three world-class alpine mountain resorts — Banff, Lake Louise and Jasper; and Banff was about an hour's bus-ride away from Calgary on a mountainous road. The conference was thoughtfully scheduled with some night sessions to allow a fair amount of day sight-seeing, because for most participants like myself, it was the chance of a lifetime to see the beautiful Canadian Rockies. A hired bus took the participants to the station of the Banff Sightseeing Gondola, which carried us to the summit of Sulphur Mountain (*ca.* 7500 ft) for a magnificent view of the city and its surroundings of picturesque mountains, meadows, lakes, and rivers. Nearer the conference venue, I saw from a distance the bright turquoise Lake Louise, and realized that its beauty befits its reputation as *'the Jewel of the Rockies.'* The Lake Louise Mountain and Ski Resort is one of N. America's biggest and reputedly the most scenic, covering both vast and varied world-class terrain in pristine wilderness. I was to learn that it is an immensely popular winter resort, attracting among overseas nationals many Japanese tourists. Karen

Overseas Conference Participation and a Bit of Travel

Crouse Badri, my former Ph.D. student, a native Canadian, had suggested to me to travel by train across the Rockies as part of my journey, but time was not on my side to allow such a luxury. Hence I could only imagine what I missed as I flew over the land decorated with snow-capped mountains and so many blue lakes and rivers. Seeing so much pristine and untapped land, though setting aside the question of inhabitability, brought to mind those overcrowded strips of land or islands elsewhere, over which there has been continual conflict.

In Florence (Italy) in 1998 for the 33rd ICCC, to which I was invited by M. Peruzzini (Italian CNR colleague), I had an experience of another cultural kind. From a professional as well as personal angle, I was very happy to meet Marcetta and Donald Darensbourg, whom I had not seen since my short stay in TAMU in 1987, and get updated on their recent research. I was also glad to meet R. Poli, my seminar chair in TAMU in 1986, and coincidentally also the chair of my session in this conference. Florence was the 'cradle of the Renaissance,' famous for its world-class art galleries and museums. A city tour took me to view Galileo's first telescope in the Science Museum, Michelangelo's famous sculpture, the marble statue of 'David' in the Academia Gallery, Italian paintings in the Uffizi Gallery, and sculptures in the Bargello Museum. There was so much for visual enjoyment, and walking around outside the museums and galleries, one could not help but admire the beautiful setting along the River Arno and the artistic bridges, especially Ponte Vecchio, the oldest and most famous of them, which uniquely houses shops selling art pieces, precious metal jewellery and souvenirs for tourists.

Nearer home, I participated in the 6th EURASIA Conference on Chemical Sciences in 2000 in Brunei. I had been in the capital Bandar Seri Bagawan in 1998, at the invitation of the Chemistry Department of Universiti Brunei Darussalam and Professor K. H. Khoo, a former colleague at UM, to give a research seminar and a public lecture of the Brunei Institute of Chemistry to a small Chemistry community. A simple hassle-free lifestyle in a clean atmospheric environment is certainly very attractive to the five hundred thousand inhabitants of this nation. I was surprised to meet several Russian scientists at this conference. Unexpectedly, one of these delegates, V. I. Sokolov (a member of the Russian Academy of Science) and I found a common interest in a study of spin-adduct formation between a chromium radical species of my interest with C_{60} (see back cover page), that famous fullerene of his interest and synthesised abundantly in his institute. This chance link led to an invitation to participate in the Mark Vol'pin 80th Anniversary Memorial International Symposium, held in the A. N. Nesmeyanov Institute of Organoelement Compounds of Russian Academy of Science, Moscow in 2003. Vol'pin was a distinguished prolific Russian chemist, an influential national scientist of his era, noted for many landmark firsts in various areas of chemistry; in particular I recall the excitement among UCL inorganic chemists when he reported for the first time 'the fixation of molecular nitrogen at transition metals' in 1965–66.

After the previous experience in my trip to Gera, I was not courageous enough to venture on my own to Moscow in a non-English speaking country. Luckily, SH was only too happy to accompany me to a country, which until lately was out of reach to most of us. As it turned out, SH's command of written Russian was sufficient to facilitate, to a certain extent, our getting around.

At the symposium, I was one of two Asian women, the other being Kazuko Matsumoto of Waseda University, Tokyo. She was one of those authors, whose elegant work I had admired and had cited often (while I was working on ruthenium-sulphur chemistry), but I did not connect 'those papers' and the presence of their author at this conference till after the event, because I was not able to infer the gender implied in a Japanese name such as Kazuko! In this respect, I must also relate another incident. Present at the Symposium was a Russian author in my area of research; we recognized each others' names, and he commented (in front of SH) 'I did not know you are a woman, why — you write so beautifully!' I did not quite know what to make of this comment; I took it as a compliment, but at the same time I could not help but think it showed that there was still a certain lack of esteem for women researchers or academics. R. Poli was also one of the participants, and I had the occasion to be updated on an area of his wide research interests. Perhaps this recent encounter was what led him to invite me to contribute an article to a special issue of the Journal of Organometallic Chemistry (v 692, 2007) on '*One-electron Reactivity in Organometallic Chemistry*', of which he was guest editor.

The symposium programme allowed for opportunities to visit the Kremlin, a must-visit in Moscow. Founded as a medieval fortress, the Kremlin is surrounded by very distinctive red-brick walls with intermittent towers of different architectural styles. It has been built upon by rulers over 800 years up to the Presidents of the Russian Federation. Within the walls are found citadels, churches and cathedrals, museums, palaces and buildings of quite an architectural eclectic mix, connected by squares and streets. Some of these buildings house the official residence of the President and the State function halls. Only one-third of the citadel territory is open to the public; even then the treasures exhibited require several days to view. In fact, the museum is one of the largest and most interesting in the world. It was regrettable that I only found time to 'float through'. It was a great disappointment to find that all performances in the world-renowned Bolshoi Ballet and Opera Theatre were fully booked daily.

May 2003 was the time of celebration of the 300th anniversary of the founding of St. Petersburg by Peter the

Great, a city named in honour of St. Peter, the apostle. It is really an uniquely beautiful city which had impressed me the most so far, and certainly compensated for the almost insurmountable hassles I encountered at the Russian Embassy in Singapore for post-conference visa extension of four days. We flew from Moscow to Pulkovo-2 International Airport, where a kindly English-speaking Russian lady, a teacher in higher education and a former visitor to East Asia, gave us bus route directions to Park Inn Pribaltiyskaya Hotel, a large four-star popular hotel on the shore of the Gulf of Finland, yet within easy access to most centres of interest of the city. It was the hotel which gave us accommodation confirmation, a prerequisite for a Russian visa extension. St. Petersburg is a unique city of numerous islands in the Gulf of Finland in the delta of the Neva River. Some observers had called it *'The City of 101 Islands'* and others *'The Venice of the North'*, referring to its many rivers and canals. The islands are connected by 500 bridges, many very beautiful and most with central sections which can be raised to allow passage of ships (for several hours in the dead of night). So much history and culture abounds here that a month's stay can easily be fully engaged. We were informed that it will take years to see the entire collection of over 3 million exhibits in the Hermitage Museum (containing the private collections of Peter the Great, Catherine the Great and expanded on by subsequent Tsars). In the few exhibition rooms that we had time to view, what impressed me deeply were the multi-coloured marbles in all colours of the rainbow, out of which the colossal statues, columns, furniture and other objects were made. With three days at our disposal, we behaved like eager over-enthusiastic tourists, trying to pack in as many of the numerous world-famous attractions as possible. A city tour, guided by a lady Engineering graduate-turned tour guide because of a recession during the time of her graduation, gave us an overview of the breath-taking historical landmarks, notably the Palace Square with the monstrous Alexander column of a single monolith of red granite, and monuments commemorating the Greats of Russia (Peter the Great, Catherine the Great, and other Tsars), and the war victories over Turkey and Napolean. We managed to see some of the city's 'must-visits', which included the Winter Palace, home of the Hermitage Museum and the residence of the Tsars; St. Isaac's Cathedral, which took 40 years to build and Kazan Cathedral; Peter and Paul Fortress. A day was taken up with a visit to Tsarskoe Selo imperial estate in Pushkin, 25 km from St. Petersburg. There I wanted to visit the Amber Room (with walls of amber) in The Catherine Palace, the town's central attraction. I was really dazzled by the exuberantly rich decor of the interiors of the palace, the extravagance of which is reported to outshine that of Versailles, which I had visited in the 1960s when a student in London. Missing a Bolshoi ballet in Moscow, we were fortunate to enjoy two evenings of ballet in the equally, if not more, renowned Marinsky Opera & Ballet Theater. Truly, there is so much interesting history and culture to absorb that St. Petersburg certainly warrants another visit of longer duration when I can!

Having briefly feasted my eyes on the rich cultural heritage of Russia, I could not help but wonder: can richness of a culture and greatness of a people only be born out of social turbulence and eventful revolutions, just like inventions out of necessity? In addition to those great monuments, cathedrals and museums so visibly displayed in Moscow and St. Petersburg, there is in addition Peter Tchaikovsky's music, Leo Tolstoy's *War and Peace*, and *Anna Karenina*, Boris Pasternak's *Dr. Zhivago* (1958 Nobel Prize, Literature), and as I had observed at the Mark Vol'pin Symposium, the wealth of Russian Chemistry, as exemplified in the work of Vol'pin, except perhaps not well-perceived in Western countries, owing to a language barrier. Riding on a sombre thought emerges a query: will it be possible for small Asian nations to support their art and science in order to catch up with the lead of a century of Western achievements? In the age of the internet, perhaps it is possible to dream for that day to come for science.

CHAPTER 13

In Service of The Profession

The years passed by rapidly and several professional activities were completed on course, such as serving on the subcommittee for Chemistry Terminology in Bahasa Malaysia (Malay), as external examiner for B.Sc. London External Examinations (Practical Chemistry, 1973–1980) and for the Graduate School, Universiti Pertanian Malaysia (1988–1995), as an elected Science Faculty representative to the Senate of UM (1982–1990), as national representative on IUPAC Commission on the Nomenclature of Inorganic Chemistry (1983–1987, and again presently), as a representative of the Malaysian Institute of Chemistry (IKM) in a seminar on 'The Role of Women in Higher Education. Implications for Higher Education in South-East Asia,' organized by the Ministry of Education of Malaysia and the Association of Southeast Asian Institute of Higher Learning (1975), and as a consultant to local firms on timber preservation and phosphate fertilizers (1980–1985). In the Malaysia Section of the Royal Society of Chemistry London, I was a committee member (1987–1989), and chairperson (1989–1991). In IKM, I had served as Inorganic Chemistry examiner for its Licentiate Membership (1985–1994), and on occasions been a member of its Council, and of its committees for Publications and Conference Organizations, and continually as referee for its journal, *Kimia,* a Malaysian Journal of Chemistry. In 2001, I received the Gold Medal of the Institute (given annually to long-term members in recognition of 'chemistry' contribution to the nation), and also the Gold Medal of Rotary Club Malaysia Research Foundation Awards in 2003. Lately I was elected a Fellow of the Malaysian Academy of Science. The IKM award carries with it a cash component in addition to the medal. As my husband SH had done before me, I had matched twice the cash value and donated to the IKM branch in my native silver state of Perak for the promotion of Chemistry. That branch had utilised the money to set up the biannual 'The Professor Goh Lai Yoong Challenge Trophy Perak Inter-school Form Six Chemistry Quiz', the inaugural one of which was held in 2002 in my Alma Mater CHIJ. The recent such occasion, held last February in Yuk Choy Secondary School, gave me an opportunity to give a talk on *'Challenges and Opportunities in Chemistry'*, with some highlights on availability of courses in NTU. It is gratifying to note that similar happenings of the last few years had helped to raise the level of interest in Chemistry among the young students in Perak.

Professor Goh Lai Yoong Challenge Trophy
Perak Inter-School Form 6 Chemistry Quiz

IKM (Perak Branch) members and teachers of participating schools, 2002

Winners and runner-ups, 2009

CHAPTER 14

My First Retirement and a Doctor of Science

Around mid-1990s, as it was mandatory to retire at age 55, I left the University of Malaya after twenty seven years of service. In that span of time, the staff establishment in Chemistry Department had doubled to over forty, with post-1986 headship passing from K. C. Chan, through V. G. K. Das to C. C. Ho. During that time, it was customary for UM to give *all* its staff (academic and non-academic) a long service award in the form of a gift after twenty years of service. At the same time it was also customary for most academic staff in the Chemistry Department to leave quietly after a lifetime of loyal service, without a farewell gesture/note of appreciation from the department or the university leadership. There had been extremes of reactions from some retired staff — from one who would not step back into the department to another who stayed on as he could flexibly blend in with the system. Perhaps many did not realise that the retirement *identity card* prepared for them by the University Security Office was that token of appreciation. But I had reasoned that at 35 years after its rebirth in 1959 as the branch in KL, UM needed more time to acquire the finesse of maturity that I had observed in century-old institutions, like Cornell or TAMU. It is of my opinion that a matter of concern to UM need not be the sentiment of some retired academics, but rather winning back the goodwill and loyalty of the forgotten alumni, the treasured human assets of most foreign institutions of today.

Likewise, my husband also retired soon after and took up a position as a senior fellow in the Department of Chemistry in NUS. In my case, I had prior arrangements with the RSC of ANU for a six months' visiting research fellowship for the purpose of writing up manuscripts. That was to be followed by an appointment as senior research fellow in Monash University in Melbourne, which suited me well as my three children were in Melbourne then — in the University of Melbourne and Monash University, and I could live with them in a house in Mt. Waverley. Thus the period January 1996 — July 1997 was spent in the Chemistry Department of Monash, working with research fellow Dr. Sue Jenkins, on an university-industry collaborative project 'Electric Windows' of Professors G. B. Deacon, L. Spiccia and D. MacFarlane. The project developed dye-sensitised solar cell technology with an industry partner, Sustainable Technologies Australia. At the end of the project time-line, still more R & D was recommended before commercialization.

While in the RSC, Martin Bennett suggested to me to submit for the D.Sc. of the University of London. Kind and helpful as he has always been, he looked up for me the format of submitted D.Sc. theses in Senate House Library during his visit to London in early 1996. Subsequently he also vetted the Introduction in my dissertation, *'Aspects of the Chemistry of Transition Metals and Main Group Elements'*, which was submitted in January 1997 from Monash University. I received the good news of success five months later. On reflection, I was proud that most of the research

My First Retirement and a Doctor of Science

was achieved while in service at UM, an institution with less than expected research infrastructure of most first world universities. Yet I am grateful for the research support it could provide. Thus, it happened that mine was the first *Malaysian* Chemistry D.Sc. conferred by a British university, though UM missed this event, perhaps until a local daily commented (quite aptly) *'the first recipient of a Doctor of Science was still active but had to leave Malaysia to work overseas to continue research and teaching'*. I attended the convocation ceremony in London on September 4, 1997, after I had taken up an appointment in NUS. As SH could not accompany me for the occasion, my sister Lai Kam and her husband James from Brisbane arranged an European trip to coincide. I also had the privilege and happiness of having my Ph.D. supervisor M. Johnson witness the ceremony and join me in the university reception that followed. That was also the period of Princess Diana's tragic passing; saddened like British citizens, I mingled with the crowd in paying respects and found solace in watching the funeral live on TV. With my sister and James, I later took a tour of Bath, saw the famous Roman baths, then went on to see Stonehenge, one of Britain's most popular attractions. Like everyone else, I stood in awe, wondering and speculating about how those gigantic stone blocks got transported to the site, in the years between 3000 and 600 BC. Certainly if such mammoth edifices could be crafted by primitive tools, and moved without transport vehicles, man will be capable of greater things than just walking on the moon!

It was timely for me to couple the D.Sc. event with a research seminar under sponsorship of the Royal Society of Chemistry Authors Travelling grant at the University of Hull (hosting department: Chemistry, with HOD Professor Ekkehardt Sinn, who is now in the Western Michigan University) on *'Homo- and Hetero-polyatomic Molecules of Groups 15 & 16 Elements in Cyclopentadienylchromium*

Chemistry', September 16, 1997. Professor Sinn was X-ray crystallography collaborator in many of my papers in the 1980–90s. It was a privilege to be hosted in his department as well as in his home and be driven to visit historic York. I had given a seminar on related aspects of the same topic, at UCL Chemistry Department (my Alma Mater) on September 5, after conferment of the D.Sc. Degree. That was initiated by M. Johnson, who though optionally retired from Chemistry in the late 1980s to become a full-time playwright, holds Emeritus Reader status in the Chemistry Department. I gathered that all former (retired, including non-Emeritus) faculty were welcome to daily coffee sessions at the department. Indeed, the academic staff of UCL Chemistry are well remembered and in today's UCL Chemistry website (http://www.chem.ucl.ac.uk/resources/history/people/index.html) is found '*A Periodic Table of Lecturers*', which provides links to all past and present faculty (with their photos), who had served in that Department. I am impressed by this record of all who had contributed to the department. In my home country it is not unusual to find institutions and departments failing to record their history for future generations, remembering the forefathers so to speak, and to allow '*out of sight*' to translate into '*out of mind*'! On this seminar occasion, I was happy to meet (for the first time) two well-known organometallic chemists, Professors A. J. Deeming and G. Hogarth. The latter in his article in 'Transition metal Dithiocarbamates', (Ch. 2, Progress in Inorganic Chemistry, v. 53, Wiley Interscience, 2005), had covered my work on dithiocarbamato chemistry done in NUS. Getting a glimpse of new exciting work of a much younger chemistry colleague was invigorating for the mind of an older member like myself. On the social side, I had an opportunity to meet up with Fiona Johnson, who had retired from teaching, then lived in the country but made it a point to travel to London for a lunch meeting.

My First Retirement and a Doctor of Science

Ph.D. and post-Ph.D.

Ph.D., 1967, at UCL

D.Sc., 1997, at UCL
(With M. D. Johnson, Ph.D. supervisor)

With ASM Honorary Fellow, Nobel
Laureate Ahmad Zewail, 2007

Passion for Science

With R. L. Huang & H. H. Lee
(M.Sc. supervisors)
and Mrs. Lee, ca. 2000

CHAPTER 15

A Second Retirement and Beyond

With a D.Sc. degree in hand and the support of then Dean of Science at NUS, Lee Soo Ying, I was offered a contract position as senior fellow in the Chemistry Department. I assumed duty in early July 1997, but my short-term stay there lasted as many as ten years, during which period I had good support from the Dean (E. C. Tan) and the various HODs (H. Chan, Y. H. Lai, H. K. Lee and A. T. S. Hor). It had been a busy and satisfying decade, involved in full-time teaching, examination of theses of graduate students, and research in a well-funded institution. There I had observed a scholarly attitude among a core group of students, who were diligently involved in research projects. The other extreme was a small non-core group, who were apparently in for Science and Chemistry not of their choice but as a last resort and hence would vent their frustrations at feedback opportunities. With moderate and adequate research funding from the Faculty and good infrastructural support from the Department, I had been able to guide and train a fair number of Honours and graduate students in research. It enabled me to resume research in organochromium chemistry, which further illustrated the wealth of reactivity patterns between specific radical species and various substrates, yielding compounds which possess yet more complexity and beauty in structural designs, illustrated in two examples below.

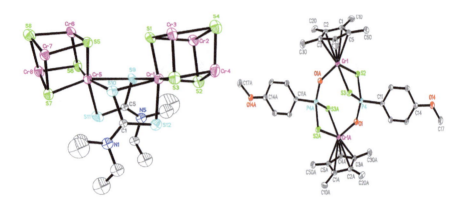

With more research students, I ventured into the chemistry of organoruthenium with sulphur ligands. This study also illustrated a wealth of new reactivity and structural features.

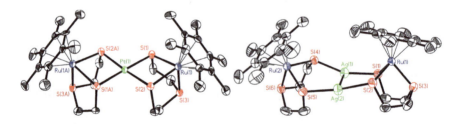

In fact, in the absence of my personal children-minding commitments and with sufficient research support, coupled with collaborations with numerous colleagues in NUS (W. K. Leong, L. L. Koh, J. J. Vittal, P. H. Leung, A. T. S. Hor, H. K. J. Yip and W. Y. Fan), and R. D. Webster (RSC of ANU), that period had been more fruitful research-wise than my whole tenure in UM. Towards the end of that period,

A Second Retirement and Beyond

I accepted with much appreciation the honour of a symposium under my name, initiated by Andy T. S. Hor, within the International Symposia ISCIC-6 and ISCOC-9, December 2006. I am very grateful to the chair of the day's symposium, H. K. Lee and the participating speakers who had graced the occasion, namely the following professors: T. C. W. Mak (Hong Kong), who had co-published with me on X-ray crystal structures, W. F. Liaw (Taiwan), the promising graduate student in M. Y. Darensbourg's group during my sabbatical in 1986, L. F. Lindoy (Sydney) whom I had met on many an occasion in Australia, S. Komiya (Tokyo), an associate in Bennett's laboratory on one of my visits there, E. Hahn (Münster), an eminent and frequent visitor at NUS Chemistry Department, H. K. J. Yip (NUS), supportive colleague and co-alumnus of HKU and R. C. S. Shin, my former Ph.D. student. The occasion also created opportunities for social interactions/exchange with other members of the international chemistry community, notably I. Reedijk (Leiden), permanent secretary of the ICCC's, Sunney I. Chan (CALTECH), academic Vice President of Academia Sinica Taiwan), G.-X. Jin (Fudan), whom I first met at the 27th ICCC in Nanjing), X.-Z. You (Nanjing), whom I also first met at the 27th ICCC and later when he was visiting professor at NUS), H. N. C. Wong (Hong Kong), co-alumnus of UCL, H. Vahrenkamp (Freiburg), whom I had cited in several of my papers, and the late R. Bau (co-alumnus of HKU). The commemorative dinner was a memorable occasion with so many well-wishers from among my international and local colleagues, and former students, with a delightful DVD presentation by my group of NUS students, prepared by S. L. Kuan, H. F. Lau and J. Tan. Thus passed my second retirement, but yet another door was to open for me.

Follow the Chemistry: Lure, Lore and Life

In NUS

SICC-2, 2004, with SH, I. Haiduc, G. B. Deacon and graduate students

Students' graduation, 2001

Commemorative Symposium Dinner, December 2006

SH, K. C. Badri, C. K. Chu

L. F. Lindoy

A Second Retirement and Beyond

My mother, still mentally alert in her 90s, who forty-five years ago had reluctantly approved of my pursuit of the Ph.D. degree, now bemoaned that she had given up hope of ever seeing me *'stop working'* in her life-time. In a way, this often fills me with some remorse, as I cannot see her or attend to her as much as I should in KL, and furthermore, I have no excuse as there really is no necessity anymore for me to *'work for a living'*, as my lifestyle is simple and inexpensive, imbued from childhood days. But perhaps I need to *'work to live'*. If one door closes and another one opens, one should move in and keep running the race and chase the dream. A time will come when one can by choice happily say that I have finished the race. A close workaholic friend advised that like him in time to come, I should 'drop with the job' and during a time of being already on the wrong side of sixty he would often marvel that it is not just scholarship and passion but 'drive, stamina and perseverance' that are needed to keep going. Yet in my case, the mind was still active and I moved again to the western tip of this island, to this rapidly expanding campus, NTU, where perhaps I am welcome as still the only female to grace the faculty of the CBC division of SPMS. So I have been able to contribute in some small ways in teaching and research.

CHAPTER

16

Family

Most importantly, I have to relate a little about a private family. I am very fortunate to have a supportive husband, who as a chemistry researcher himself understands the demands of the profession. I have heard that some of my women colleagues in UM had to observe 9 to 5 office hours, as expected by their husbands or for other reasons; it was no wonder some of them complained that they could not have fulfilling careers. Many even exclaimed that as professor I should not need to keep long hours at work.

Like many Chinese extended families, my siblings and I have benefited from the close-knit families of our parents and their relatives, who had helped unconditionally in times of our need. Coming from a family with only a modest income, it was a wonder that my siblings and I all secured higher education. That was made possible by the endowment of scholarships for deserving students. Two of my sisters were awarded Federal scholarships for courses in Science and Pharmacy, respectively, in UM. My two brothers were supported on parental scholarships and graduated with MBBS degrees from UM, at the (fortunately low) cost of 600 Malayan dollars for annual fees, which remained highly subsidised by the state till today (about RM 2000 per annum).

These expenditures, in addition to that for my third sister in Dublin, exhausted my parents' resources, even though I as the first-born felt obligated to contribute a little by temporary school teaching during my university summer vacations.

We have been blessed with three well-behaved and highly independent children. As I observed the after-school tuition demands on modern day parents, I am thankful that we never had that concern. Undoubtedly, good teachers in reputable schools made all the difference. To give them a balanced education, we had organised for them extracurricular activities, such as piano and violin, gymnastics, Tae-kwan-do and swimming. Our choices of institutions for spending of our sabbatical leaves were made with their education exposure in mind, and we were glad that our stays in RSC/ANU in Canberra and TAMU in College Station had turned out very beneficial to them. After our sabbatical in Canberra in 1991 and noting their independent capability, we had decided that son CC and daughter SY continue their pre-university years (Australian Years 11 and 12) in Xavier's College and Methodist Girls' College (MLC), two very reputable schools in Kew, Melbourne, with my sister Lai Fong (then settled in Melbourne) as guardian. That would give them full immersion in the English medium of instruction as well as the Australian syllabus. The youngest, SL, was to go to MLC, two years later after Form 3 in Malaysia. Thus, my frequent visits to RSC/ANU and Monash during the years 1991–94 gave me opportunities to visit the children as well. Likewise, the two years 1995–97 spent in Australia after retirement from UM was equally timely. I could make a home for them in Mt. Waverley, a suburb somewhat mid-way between Monash University in Clayton and University of Melbourne in the city, where they were studying.

Family

With parents and siblings, ca. 1952

With parents, B.Sc., HKU, 1963

Graduation of SL (the youngest), Monash University, 1999

While I was in NUS, the children graduated: CC (MBBS, Monash 1998; MBA, Monash 2006; MRACMA, 2006), SL (Bachelor of Commerce — Accounting/Econometrics, Hons H2A, Monash 1999) and SY (MBBS Hons 1 and BMedSc, Melbourne 2000; FACD, 2008). SY has since obtained her specialisation (Dermatology, Melbourne, 2008). When they were young, our children had wondered why their parents had 'homework' to do at night and perhaps already then, came to the decision that careers might be more satisfying as medical practitioners like their maternal uncles who did not have to bring home work at night. In their eyes they must have seen their chosen vocations as befitting their times while as parents we may be pleased of their success, yet could not help but wonder why we were not role models as professional scientists. While academics may have many advantages such as flexible work hours, excitement of research and discovery and ability to plan for overseas visits and conferences, the present day choices are more focused on monetary values and avoidance of the excessive competition as present in academia. Except for the brave, the recent criteria for employment in good universities can be quite intimidating to some, e.g. scholarship, pedigree degrees, outstanding research and potential for good teaching. Spending unending hours in the laboratory and/or library or at the computer to solve problems of science, may not likely be the choice of the average graduate. Nevertheless, they would have their last say that 'if one cannot do, one can always teach'. The world has changed; nowadays by the time students enrol at the university they have plenty of opportunities and vocational choices — professions, research, academia, business startups, sales and the like.

Family

At Leisure

Niagara Falls, 1967

Stonehenge, UK, 1997

Campus of the Chinese University of Hong Kong, 2004

Jiu ZhaiGou, China, 2005

Hope City, B.C., Canada, 2007

Trivia

This little information, though a little ordinary and boring, may be of interest to some people. Being brought up frugal like my parents due to the circumstances described previously, I remain a careful and financially sound saver, even though I carry a few credit cards automatically given by the banks. For dresses and costumes, many pieces were made by an aunt with expert tailoring skills before her retirement; otherwise clothes and shoes are bought as when needed, more for comfort than for brand names. As it is reasoned that the profession does not demand glamour, I have not indulged in cosmetics and stylized dressing. Also, perhaps not unlike most academics who may be described mostly as not financially savvy, minor assets accumulated include only a place to live in (semi-detached bungalow in Malaysia) and a couple of others 'to live on'. Minor small investments in equities, gold and funds remain useful but these are at the mercy of the health of banks. As described previously, games and hobbies were few with sewing being a favourite, and leisure times were mostly spent on travel or with family and friends or perhaps for a musical/ballet which happened to come to town. A potential weakness may include a liking for probability games but exposure to these potentially fatal attractions had been very low as they have not been readily accessible over the years. An incompetent driver, I would rather be chauffeured around, having once poorly parked our Mercedes 200 which was then to be battered by a bad lorry driver; but fortunately there was little feeling of pain for repairing a car which was already a decade old.

Epilogue — Reflections

Forty years in academia is an era of a sort; there had been trying, as well as, satisfying times. A discomforting realisation is that more could have been produced research-wise. A regret was that a good capable Ph.D. candidate Othman Nor, under my supervision upon the departure of his original supervisor, did not submit his thesis by the time I left UM. In essence the experimental work was completed years earlier, but unfortunately he was called upon to shoulder administrative duties with positions of Director of UM's Centre of Foundation Studies in Science, followed by Deputy Vice-Chancellor of research. Happiness was when I witnessed students graduate with their first degrees, and graduate students finish with their Master's or Ph.D.'s. Interaction with students has been mentally invigorating as it is always refreshing and full of hope, new every year. It is more gratifying to meet them in later years in variously important positions in different areas of science-related or business endeavours. A person in point is Lee Soo Ying who has risen to a leadership role in tertiary education, and Lam Yoke Ching, a female Honours graduate, is among many others who had risen to directorship roles in The Department of Chemistry (Malaysia) (a body of a hundred years' standing, now under The Ministry of Science, Technology and Innovation, regulating matters on Forensic Science, Environmental Health, Research and Quality Assurance, Industry and Classification Trade Tariff, and Development and Information Technology). Many of these ex-students have seen their early retirements as well, and as such the loss of experienced leading personnel in such bodies, as in academia, constitutes a serious loss of manpower expertise in the nation. It was only recently that Malaysia realized that its experts were retired prematurely, so that presently experienced and productive staff in academia and in the civil service are given service extensions, though only for a meagre one to three years period.

Malaysia has gone through turbulent times in university education. First in the early 1980s, the use of the national language Malay as teaching medium (introduced in early 1970s in schools) was extended to university education. However, there were no textbooks in Malay for the higher years, and translations were slow and might not be up to date; often lecturers got by with impromptu coining of new chemical terminology. It had been unwise to create terminology when there was none and even worse to change nomenclature. The students were naturally disadvantaged and that was partly the start of the setback for the country's scientific education and progress. It is only when a country gets close to a developed status with a need to compete in research and technology to produce new science, new products and services that the need for technical and business English would be much appreciated, in view of vast bodies of accumulated knowledge in English. Next was the explosive expansion of higher education for the masses. Partly in response to this, UM initiated a sort of 'training scheme' to help increase the number of Ph.D. holders among its academic staff; these were to be trained mainly in universities in UK. As a result, recruitment criteria in UM then and now, as also in other public universities in

Malaysia, became much less stringent than in most world-class universities. In these universities, potential recruits into their teaching faculty are selected from the world-wide pool and may need at least one postdoctoral experience of two years in a premier university. This is to ensure that the new faculty already has gained valuable experience from a research-intensive centre, has demonstrated evidence of independent research capability, and is ready to lead in new research directions.

Reflecting on the early days of my return to teach at UM, it was clear that there was a vast chasm between the developed world and Malaysia, naturally translating down to research endeavours. Whereas academic research had raced ahead to solve issues in the developed world, the local needs were more mundane, such as for adaptation of analytical methods for local industries or less challenging problems of the tin and rubber industries (and later palm oil) which already had their own research institutes. As funding had largely been allocated to these institutes, the funding for any other areas of research was minimal. However, determined researchers managed to find their little niches and means if they were not being overly ambitious. It is only later that in anticipation of the country's needs for original products and services rather than depending on imported ones, that funding came in a big way around late 1980s. However, research cannot be promulgated with sudden inputs of funds; in research one has to keep running at the forefront all the time or otherwise drop out of the picture. It is not something that can be turned on rapidly or suddenly when a need arises e.g. to sustain development when a country gets to be on the verge of a developed status. Hence there may have been apparent wastage of funds. In many ways the early scientific funding provided the first opportunity, an experiment perhaps, for many staff of the universities and institutes to just get acquainted with research culture and hopefully later years would bring better results. Having invested in improving research infrastructure in the universities, there is only need to follow up by establishing a granting system as practised in developed countries, which incorporates proper peer reviews by active researchers at both local and international levels of grant applicants (for track records) and grant proposals (for feasibility of the projects), and finally a serious assessment of the outcomes of those projects. Further, later assessments have to be made for the potential of the results and an evaluation of accountability where necessary. Once the research culture is in place and good and original science and genuine scientists are allowed to flourish and be rewarded, hopefully the universal ranking of Malaysian universities will rise in tandem with generous research funding provided, with a few percent of GDP spent on science, a necessity in this modern age. As one of its ex-professors, it is indeed heartening to note that ranking of UM has improved to above 200 in the just-released 2009 Times Higher *Education-QS World University Rankings*, and I reckon that higher rankings will follow with present efforts at improving and encouraging quality in staff and students and their outputs.

Sabbatical leaves were essential and had allowed me to focus on research for a few months and refresh the mind. These together with conference participation provided an opportunity to keep up at the forefront, as the landscape was always changing, e.g. directions of research, new instrumental techniques and computer capabilities. Furthermore, spending these leaves overseas had enabled me to forge links with prominent professors in top-notch departments, giving me renewed inspiration from the research vibrancy of the place, as well as an opportunity to experience an exchange of hands-on bench practices in difficult manipulations with the post-doctoral fellows, who had come from world-class research environments, say, in Germany, US, UK and Japan. I had attempted to transfer any newly-acquired

Epilogue — Reflections

techniques to my research students and I trust these are carried on to next generation chemists by academics who had gone through my training, *viz.* recently-retired Professor Karen C. Badri of Universiti Putra Malaysia and Associate Professor R. C. S. Wong of UM.

Taking the children along with us for sabbatical leaves overseas gave me peace of mind as I worked in the laboratory or library, and definitely also exposed their young minds to new horizons. We were very appreciative of the support from Kaye Beckwith (1981) who saw to the needs of our family from arrival at the airport in Canberra till we were settled in ANU housing. In College Station, Marcetta Darensbourg had organised for the members of the Malaysian Students Association to drive us in two cars from the airport, a two hours' journey to College Station.

International conferences are of special relevance, as they provide a forum for exposure of one's work and indirectly of the affiliated institution, a good survey of work at the forefront in a short time and an opportunity to discuss/interact with researchers in the same or related fields, and collaborative outcomes have sometimes ensued. For instance, as mentioned above, a small EURASIA conference in Brunei in 2000 gave me an opportunity to interact with Russian chemists, unlikely visitors to this Asian part of the world. At the Inorganic Chemistry Conference of the RACI in Melbourne in 2003, Professor R. Boére (University of Lethbridge, Canada) and I found common ground to initiate reactions between organometallic and main group radical species, which had resulted in the realisation of a few much-sought-after unlikely molecules and a couple of high-impact journal papers. We had missed each other at 'IRIS VII' in Banff in 1994; otherwise collaborative research would have started much earlier. I was glad to meet him again at the 90th Canadian Chemistry Conference in Winnipeg in 2007. It is not only from fellow scientists but the new and varied sights and sounds from foreign lands that continue to add inspiration to research or to reveal new perspectives towards life.

Malaysian academia has a fair quota of women members, 20–25% of an establishment of more than forty members in Chemistry in UM when I left, and currently 33%. Similarly in Singapore, I found myself one of approximately 20% women in NUS Chemistry Department. These percentages are strikingly lower in developed countries notably USA. One only hopes that the lower attrition rate of female academics in less developed countries does not correlate with less challenging demands in research, making universities much like 'glorified' high schools. While the numbers are still far from the 'natural' 50%, the question had always been asked why or why not, considering that in our undergraduate classes and in graduate degrees, the girls often outnumber the boys. So why are they not in academia? Have girls found the Ph.D. and research too challenging, or do they envisage a greater challenge beyond, or already experience 'hitting a glass ceiling' early in the climb? Even in first world countries, where there are action guidelines in place to encourage women (a minority group), there are few that I know of the calibre of Jacqueline Barton (CALTECH), Carolyn Bertozzii (UC Berkeley), Marcetta Y. Darensbourg and Kim R. Dunbar (both of TAMU), Joan S. Valentine (UCLA), Marianne Baudler (Köln, Germany), an eminent phosphorus chemist, who was research-active till her 70's, and in Asia V. W. W. Yam and D. Yang (HKU). Perhaps the long education process to the Ph.D. and postdoctorate may be the deterrent to raising families. Do present day young chemists take as role models, such as Kathleen Lonsdale, Dorothy Hodgkin and the exceptional Nobel Laureate models Marie Sklodowska-Curie and Irene Joliot-Curie, who had all successfully combined family and career? A local role model to aspiring female graduate students is Karen C. Badri of UPM, who had overcome great odds to successfully combine a full-time

Epilogue — Reflections

faculty position with bringing up a family of four children during her Ph.D. tenure.

Commenting on the same topic, female chemists today are indeed fortunate that *extreme* sexism at the work place of the intensity as found in the late 19th century is no longer existent, thanks to the efforts of male professors Augustus Harcourt, William Ramsay, William Tilden and J. D. Bernal, in their fight for the rights of women chemists. There was a time when 'many male chemists fought bitterly to prevent women joining their ranks'.* Indeed, the atmosphere of sexism persisted to the days of Rosalind Franklin, who was reported to find the atmosphere so overbearing in King's College London, that in 1953 she had to find haven in the laboratory of J. D. Bernal at Birkbeck College. Although J. Watson had dismissed her role in the discovery of the double helix in his book *The Double Helix*, Franklin is today honoured in the Rosalind Franklin University of Medicine and Science, established in 2004, comprised of four colleges and schools, one being The Chicago Medical School. The gender issue is being aired to the present day; for instance, the Editorial of Chemistry World (August 2008) noted that 'academic chemistry is a less welcoming environment for women than for men' and that the RSC and UKRC (UK Resource Centre for Women in Science, Engineering and Technology) are 'looking at how to stem the loss of female chemists.' In my time I might unknowingly have had a fair share of such and other unholy practices as it had taken such a long time to attain professorship while lesser people achieved this much earlier from other types of criteria. While there is still real concern about 'female attrition in science',[†] there is an

---

*See: Marlene and Geoff Rayner-Canham, in *'Fight for Rights'* in *Chemistry World*, March 2009, pp. 56–59.
[†]See RSC Policy Bulletin of The Royal Society of Chemistry London, *'Advancing the Chemical Sciences'*, Issue No. 10, October 2008.

encouraging observation that the number of senior female academics has been steadily rising in the last two decades. However, the most recent report[‡] concluded that "the number of women in chemistry and in senior positions is not intractable", and "Good Practice"[#] in Chemistry Departments is strongly encouraged. Nations at the forefront are realizing that they cannot afford to lose the contribution accruing from 50% of their population. Indeed, the girls should and would be encouraged and supported towards this national goal.

Finally, as we all know, only one thing is certain in life, but it is a top-of-the-world feeling to remain useful like a small candle to light up some windows for glimpses into future dreams.

[‡]RSC News — Policy Bulletin — November 2009: 'Staying in Science', (The RSC's Policy Bulletin looks at Women's Career Progression in UK Chemistry Departments).

[#]The RSC has published 2 reports: 'Good Practice of University Chemistry Departments', 2004. 'Planning for success: Good Practice in University Chemistry Departments', 2008.

APPENDIX I

List of Acronyms

ANU	Australian National University
ASM	Academy Science Malaysia
BMedSc	Bachelor of Medical Science
CALTECH	California Institute of Technology
CBC	Chemistry and Biological Chemistry
CHIJ	Convent of the Holy Infant Jesus
CNR (Italy)	National Research Council, Italy
CSLCE	Cambridge School Leaving Certificate Examination
ESR	Electron Spin Resonance
EURASIA	Europe Asia
FACD	Fellow of the Australasian College of Dermatologists
GDP	Gross domestic product
HK	Hong Kong
HKU	University of Hong Kong
HOD	Head of Department
ICCC	International Coordination Chemistry Conference
ICC	International Cricket Council
IC	Imperial College London
IC-03	Inorganic Chemistry Conference, 2003, of RACI
IKM	Institut Kimia Malaysia
IRIS	International Symposium on Inorganic Ring Systems
ISCIC-6	6th International Symposium of Chinese Inorganic Chemists
ISCOC-9	9th International Symposium of Chinese Organic Chemists
IUPAC	International Union of Pure and Applied Chemistry

JP	Justice of Peace
KL	Kuala Lumpur
LHT	Lady Ho Tung
LP	Lillian Penson
MBA	Master in Business Administration
MBBS	Bachelor of Medicine and Bachelor of Surgery
MIT	Massachusetts Institute of Technology
MRACMA	Member of Royal Australian College of Medical Administrators
MRCP	Member of the Royal College of Physicians (U.K.)
NASA	National Aeronautics and Space Agency
NIH	National Institute of Health
NMR	Nuclear Magnetic Resonance
NSF	National Science Foundation
NSW	New South Wales
NTU	Nanyang Technological University
NUS	National University of Singapore
OZOM	Australian Symposium on Organometallic Chemistry
PJ	Petaling Jaya
PWD	Public Works Department
RACI	Royal Australian Institute of Chemistry
RM	Malaysian ringgit
RRIM	Rubber Research Institute of Malaysia
RSC	Research School of Chemistry
SICC-2	Singapore International Chemical Conference-2
SMI	Saint Michael's Institution
SPMS	School of Physical and Mathematical Sciences
TAMU	Texas A&M University
UC	University of Chicago
UC Berkeley	University of California, Berkeley
UCL	University College London
UCLA	University of California, Los Angeles
UK	United Kingdom
UM	University of Malaya
UNSW	University of New South Wales
USC	University of Southern California
USSR	The Union of Soviet Socialist Republics
UTAR	University Tunku Abdul Rahman
WWII	World War Two

APPENDIX II

Selected List of Journal Publications*

2009 'A Cyclometallated (azobenzene) Palladium(II) complex of 9S3 (1,4,7-trithiacyclononane). Synthesis and reactivity with thioether-dithiolate metalloligands. Single-crystal X-ray diffraction analyses and electrochemical studies', R. Y. C. Shin, C. L. Goh, L. Y. Goh, R. D. Webster, Y. Li. *Eur. J. Inorg. Chem.* 2009, 2282–2293.

2008 'Coupling of $CpCr(CO)_3$ and Heterocyclic Dithiadiazolyl Radicals. Synthetic, X-ray diffraction, Dynamic NMR, EPR, CV and DFT Studies', H. F. Lau, P. C. Y. Ang, V. W. L. Ng, S. L. Kuan, L. Y. Goh, A. S. Borisov, P. Hazendonk, T. L. Roemmele, R. T. Boeré, R. D. Webster. *Inorg. Chem.* 2008, **47**, 632–644.

2007 'Synthetic and X-ray Structural and Reactivity Studies of Cp*Ru(IV) Complexes containing Bidentate Dithiocarbonate, xanthate, carbonate and phosphinate Ligands ($Cp^* = \eta^5\text{-}C_5Me_5$)', E. P. L. Tay, S. L. Kuan, W. K. Leong, L. Y. Goh. *Inorg. Chem.* 2007, **46**, 1440–1450.

---

*For a complete listing, refer to one of the following:
  (i) ISI, WEB of Knowledge, WEB of Science
      Author Finder: WONG L Y (1965–1970) and GOH L Y (after 1970)
 (ii) http://www.spms.ntu.edu.sg/CBC/Personal/GLY.html
(iii) http://chmgsh.tripod.com/index.htm

'Sulfur-alkylation-initiated Cp*Ru thiyl radicals',
  R. Y. C. Shin, H. S. Sim, L. Y. Goh, R. D. Webster.
  J. Organomet. Chem. 2007, **692**, 3267–3276.
'First Examples of a Dithiolato Bridge in (η^6-C_6Me_6)
  RuII-Cr(0,II,III) Complexes. Synthetic, Single-Crystal
  X-ray Diffraction, and Electrochemical Studies',
  R. Y. C. Shin, V. W. L. Ng, L. L. Koh, G. K. Tan,
  L. Y. Goh, R. D. Webster. *Organometallics* 2007, **26**,
  4555–4561.
'Tethered Indenyl-Phosphine Complexes of Ruthenium(II) via
  Reductive Elimination of a Ruthenium(IV) Complex',
  S. Y. Ng, G. K. Tan, L. L. Koh, W. K.. Leong, L. Y. Goh.
  Organometallics 2007, **26**, 3352–3361.

2006 'HMB and Cp* Ruthenium(II) Complexes containing Bis- and
  Tris-(mercaptomethimazolyl)borate Ligands: Synthetic,
  X-ray Structural and Electrochemical Studies
  (HMB = η^6-C_6Me_6, Cp* = η^5-C_5Me_5)', S. L. Kuan,
  W. K. Leong, L. Y. Goh, R. D. Webster. *J. Organomet. Chem.*
  2006, **691**, 907–915.
'Pentamethylcyclopentadienyl Ruthenium(III) vs
  Hexamethylbenzene Ruthenium(II) in Sulfur-Centered
  Reactivity of Their Thioether-Thiolate and Allied
  Complexes', R. Y. C. Shin, L. Y. Goh. *Acc. Chem. Res.*
  2006, **39**, 301–313.
'Indenyl Ruthenium Complexes Containing
  1,1'-Bis(diphenylphosphino)ferrocene (dppf) and Thiolato
  Ligands: Synthesis, X-ray Structure Analysis,
  Electrochemistry and Magnetic Studies', S. Y. Ng,
  W. K. Leong, L. Y. Goh. R. D. Webster. *Eur. J. Inorg. Chem.*
  2006, 463–471.
'Highly Oxidized Ruthenium Organometallic Compounds.
  The Synthesis and One-Electron Electrochemical
  Oxidation of [Cp*RuCl$_2$(S$_2$CR)] (Cp* = η^5-C_5Me_5, R = NMe_2,
  NEt_2, O^{i}Pr)', S. L. Kuan, E. P. L. Tay, W. K. Leong,
  L. Y. Goh, C. Y. Lin, P. M. W. Gill, R. D. Webster.
  Organometallics 2006, **25**, 6134–6141.
'Cyclopentadienylchromium complexes of
  1,2,3,5-dithiadiazolyls: η^2 π complexes of cyclic sulfur-
  nitrogen compounds', H. F. Lau, V. W. L. Ng, L. L. Koh,
  G. K. Tan, L. Y. Goh, T. L. Roemmele, S. D. Sonja,
  R. T. Boeré. *Angew. Chem. Int. Ed.* 2006, **45**, 4498–4501.

Selected List of Journal Publications

'η^1 and η^2 complexes of λ^3-1,2,4,6-thiatriazinyls with $CpCr(CO)_x$ {x = 2,3}', C. Y. Ang, R. T. Boeré, L. Y. Goh, L. L. Koh, S. L. Kuan, G. K. Tan and X. Yu. *Chem. Commun.* 2006, 4735–4737.

2005 'An Organometallic Tetranuclear Complex of μ_4-PO_4: [{(Cp^*Cr)$_2$(μ-OMe)$_2$}$_2$(μ_4-PO_4)]X (X = I, PF_6)', R. Y. C. Shin, G. K. Tan, L. L. Koh, L. Y. Goh, R. D. Webster. *Organometallics* 2005, **24**, 1401–1403.

'Reactivity of $[CpCr(CO)_3]_2$ towards thione (C=S) moieties in some sulfur-containing substrates', V. W. L. Ng, S. L. Kuan, Z. Weng, W. K. Leong, J. J. Vittal, L. L. Koh, G. K. Tan, L. Y. Goh. *J. Organomet. Chem.* 2005, **690**, 2323–2332.

'Heterocyclic Thionates as a New Class of Bridging Ligands in Oxo-centered Triangular Cyclopentadienyl-chromium(III) Complexes', V. W. L. Ng, S. L. Kuan, W. K. Leong, L. L. Koh, G. K. Tan, L. Y. Goh, R. W. Webster. *Inorg. Chem.* 2005, **44**, 5229–5240.

'(Cyclopentadienyl)chromiumtricarbonyl dimers as a source of metal-centered free-radicals to form stable η^2-bonded spin-adducts with fullerenes', V. I. Sokolov, R. G. Gasanov, L. Y. Goh, Z. Weng, A. L. Chistyakov, I. V. Stankevich. *J. Organomet. Chem.* 2005, **690**, 2333–2338.

'Metallophilicity in Annular Ru_2M_2 Derivatives of $(HMB)Ru^{II}(tpdt)$ versus (Bis)-η^2-dithiolate-bonding in Ru_2M Derivatives of Cp^*Ru^{III}(tpdt) (HMB = η^6-C_6Me_6; Cp^* = η^5-C_5Me_5; M = Cu^I, $Ag^{I,II}$, Au^I; tpdt = 3-thiapentane-1,5-dithiolate)', R. Y. C. Shin, G. K. Tan, L. L. Koh, J. J. Vittal, L. Y. Goh,* R. D. Webster. *Organometallics* 2005, **24**, 539–551.

'S-Alkylation-induced Redox Reactions leading to Reversible Sulfur-Sulfur Coupling in a Pentamethylcyclopentadienyl Ruthenium(III) Thiolate-Thioether System', R. Y. C. Shin, M. E. Teo, W. K. Leong, J. J. Vittal, J. H. K. Yip, L. Y. Goh, R. W. Webster. *Organometallics* 2005, **24**, 1483–1494.

'Square Planar versus Tetrahedral NiS_4 Cores in the Coordination Spheres of (HMB)Ru(II) and Cp^*Ru(III) and a Related CuS_4 Complex. Synthetic, Single Crystal X-ray Diffraction and Magnetic Studies {HMB = η^6-C_6Me_6 and Cp^* = η^5-C_5Me_5}', R. Y. C. Shin, M. E. Teo, L. L. Koh, G. K. Tan, J. J. Vittal, L. Y. Goh, K. S. Murray,

B. Moubaraki, X.-Y. Zhou. *Organometallics* 2005, **24**, 4265–4273.

'Redox Dependent Isomerization of Organometallic Ru^{II}/Ru^{III} Compounds Containing the Hydrotris(Methimazolyl)Borate Ligand. An Electrochemical Square Scheme Mechanism', S. L. Kuan, W. K. Leong, L. Y. Goh, R. D. Webster. *Organometallics* 2005, **24**, 4639–4648.

2004 'Homolytic Cleavage and Aggregation Processes in Cyclopentadienylchromium Chemistry', Z. Weng, L. Y. Goh. *Acc. Chem. Res.* 2004, **37**, 187–199.

'Syntheses and X-ray Crystal Structures of Di- and Tri-Nuclear Trithiolate/thioether Bridged Complexes of Ruthenium. Electrochemistry of Mixed Valence Triruthenium Complexes', R. Y. C. Shin, S. Y. Ng, G. K. Tan, L. L. Koh, S. B. Khoo, L. Y. Goh, R. W. Webster. *Organometallics* 2004, **23**, 547–558.

'[(L$_n$)Ru{η^3-(tpdt)}] Complexes as Dithiolate Donors to Group 10 Metal Centers: Synthetic, Single Crystal X-ray Diffraction and Electrochemical Studies. {L$_n$ = η^6-C$_6$Me$_6$ (HMB) and η^5-C$_5$Me$_5$ (Cp*); tpdt = S(CH$_2$CH$_2$S)$_2$}', R. Y. C. Shin, G. K. Tan, L. L. Koh, L. Y. Goh, R. W. Webster. *Organometallics* 2004, **23**, 6108–6115.

2003 'Complexes from ring opening of Lawesson's reagent and phosphorus-phosphorus coupling', Z. Weng, W. K. Leong, J. J. Vittal, L. Y. Goh. *Organometallics* 2003, **22**, 1645–1656.

'Arene-ruthenium Complexes of an Acyclic Thiolate-thioether and Tridentate Thioether Derivatives resulting from Ring-closure Reactions', R. Y. C. Shin, M. A. Bennett, L. Y. Goh, W. Chen, D. C. R. Hockless, W. K. Leong, K. Mashima, A. C. Willis. *Inorg. Chem.* 2003, **42**, 95–106.

2002 'Organometallic-initiated cleavage of the metal chelate and thiazole rings in a cyclopentadienylchromium Complex', L. Y. Goh, Z. Weng, W. K. Leong, J. J. Vittal. *J. Am. Chem. Soc.* 2002, **124**, 8804–8805.

'Organometallic radical-initiated carbon–sulfur bond cleavage and carbon–carbon coupling in dithiocarbamate and thiocarbenoid cyclopentadienylchromium complexes', L. Y. Goh, Z. Weng, T. S. A. Hor, W. K. Leong. *Organometallics* 2002, **21**, 4408–4414.

Selected List of Journal Publications

'Reactions of bis(thiophosphoryl)disulfanes and bis(thiophosphinyl)disulfanes with metal species: an alternative, convenient route to metal complex and organometallic dithiophosphates and dithiophosphinates', I. Haiduc, L. Y. Goh. *Coord. Chem. Rev.* 2002, **224**, 151–170.

2001 'C-S bond cleavage and C-C coupling in cyclopentadienyl-chromium complexes to give the first dithiooxamide-bridged and doubly dithiocarbamate-bridged double cubanes: $[Cp_6Cr_8S_8\{(C(S)NEt_2)_2\}]$ and $[Cp_6Cr_8S_8(S_2CNEt_2)_2]$', L. Y. Goh, Z. Weng, W. K. Leong, P. H. Leung. *Angew. Chem. Int. Ed.* 2001, **40**, 3236–3239.

1999 'Tetranuclear Complexes from the Reaction of Tetraphosphorus Triselenide with Cyclopentadienyl-chromium Tricarbonyl Dimer. Synthesis, Thermal Degradation and X-ray Crystal Structures of $Cp_4Cr_4(CO)_9(P_4Se_3)\cdot 1/2 C_6H_6$ and $Cp_4Cr_4(CO)_8(P_2Se_2)$', L. Y. Goh, W. Chen, R. C. S. Wong. *Organometallics* 1999, **18**, 306–314.

'Polyatomic molecules and aggregates of main group 15/16 elements in organochromium chemistry', L. Y. Goh, *Coord. Chem. Rev.* 1999, **185-186**, 257–276.

'A facile reaction of Sb_2S_3 with $[CpCr(CO)_3]_2$: formation of a novel tetrachromium complex $[CpCr(CO)_3]_4(Sb_2S)$', L. Y. Goh, Wei Chen, R. C. S. Wong. *J. Chem. Soc., Chem. Commun.* 1999, 1481–1482.

1996 'Base-Induced Fragmentation of a Macrocyclic Thioether at an (Arene)ruthenium(II) Center. Generation of η^1-(S)-Ethenethiolate and η^2-C,S-Thioacetaldehyde', M. A. Bennett, L. Y. Goh, A. C. Willis. *J. Am. Chem. Soc.* 1996, **118**, 4984–4992.

1995 'Synthesis and Structure of endo-Coordinated o-Xylylene Complexes of Zerovalent Ruthenium and Osmium, $M\{\eta^4\text{-}o\text{-}(CH_2)_2C_6H_4\}(PMe_2Ph)_3$ (M = Ru, Os), and of a Tricarbonyliron Adduct of the Ruthenium Complex', M. A. Bennett, M. Bown, L. Y. Goh, D. C. R. Hockless, T. R. B. Mitchell. *Organometallics* 1995, **14**, 1000–1007.

'A Tetrachromium Complex from the Cage-opening of P_4S_3 by Cyclopentadienylchromium Tricarbonyl. Synthesis, X-Ray Crystal Structure and Thermal Degradation of $Cp_4Cr_4(CO)_9(P_4S_3)$. $CpCr(CO)_3H$ as a Byproduct',

L. Y. Goh, W. Chen, R. C. S. Wong, K. Karaghiosoff. *Organometallics* 1995, **14**, 3886–3896.

1994 '(Tellurolato)chromium Complexes. Syntheses and Crystal Structures of CpCr(CO)$_3$(TePh), [CpCr(CO)$_2$(TePh)]$_2$, and [CpCr(TePh)]$_2$Te', L. Y. Goh, M. S. Tay, W. Chen. *Organometallics* 1994, **13**, 1813–1820.

1993 'Novel Polycyclic Phosphane-to-Metal Coordination.Reaction of [CpCr(CO)$_3$]$_2$ with Elemental Phosphorus and Structure and Paramagnetism of the Odd-Electron Complex [CpCr(CO)$_2$]$_5$P$_{10}$', L. Y. Goh, R. C. S. Wong, E. Sinn. *Organometallics* 1993, **12**, 888–894.

'Unprecedented Cage-Opening of P$_4$S$_3$ Initiated by an Organometallic Radical: Synthesis and Structure of [Cp$_4$Cr$_4$(CO)$_9$(P$_4$S$_3$)]', L. Y. Goh, W. Chen, R. C. S. Wong. *Angew Chem. Int. Ed. Engl.* 1993, **32**, 1728–1729.

1992 'Thiolate-Bridged Dichromium Complexes. Syntheses and Crystal Structures of [CpCr(CO)$_2$(SPh)]$_2$ and [CpCr(SPh)]$_2$S', L. Y. Goh, M. S. Tay, T. C. W. Mak, R.-J. Wang. *Organometallics* 1992, **11**, 1711–1717.

'η^5-Cyclopentadienylchromium Complexes of Sulphur', L. Y. Goh. *Inorg. Synth.* 1992, **29**, 251–254.

'η^5-Cyclopentadienylchromium Complexes of Phosphorus, Tetracarbonbyl-μ-η^2-diphosphido-bis (η^5-cyclopentadienylchromium), [CpCr(CO)$_2$]$_2$(μ-η^2-P$_2$) and Dicarbonylcyclotriphosphido-(η^5-cyclopentadienylchromium), CpCr(CO)$_2$(η^3-P$_3$)', L. Y. Goh, R. S. C. Wong. *Inorg. Synth.* 1992, **29**, 247–250.

'Double Deprotonation, Ring-Opening and C-C Bond Formation in a Coordinated Crown Thioether. Formation and Structure of an Areneruthenium(II) Ethenethiolate, [Ru(SCH=CH$_2$)(η^6-C$_6$Me$_5$CH$_2$CH$_2$CH$_2$SCH$_2$CH$_2$S)]', M. A. Bennett, L. Y. Goh, A. C. Willis. *J. Chem. Soc., Chem. Commun.* 1992, 1180–1182.

1991 'Determination of metal-metal bond energies in [CpCr(CO)$_2$L]$_2$ complexes [L = CO, P(OMe)$_3$] by NMR', L. Y. Goh, Y. Y. Lim. *J. Organomet. Chem.* 1991, **402**, 209–214.

1990 'Reaction of [{Cr(cp)(CO)$_3$}$_2$] (cp = η^5-C$_5$H$_5$) with Elemental Phosphorus. Isolation of Cr$_2$(cp)$_5$P$_5$ as a Thermolysis Product and its X-Ray Crystal Structure', L. Y. Goh,

R. C. S. Wong, C. K. Chu, T. W. Hambley. *J. Chem. Soc., Dalton Trans.* 1990, 977–982.

'A New Type of Polycyclophosphidochromium Cluster, [(Cp)Cr(CO)$_2$]$_5$P$_{10}$ (Cp = η^5-C$_5$H$_5$). First Observation of Polycyclic P-to-M Coordination', L. Y. Goh, R. C. S. Wong, E. Sinn. *J. Chem. Soc., Chem. Commun.* 1990, 1484–1485.

'Equilibrium studies on the metal-metal bond dissociation in the [(C$_5$Me$_5$)Cr(CO)$_3$]$_2$ dimer by NMR spectroscopy', L. Y. Goh, S. K. Khoo, Y. Y. Lim. *J. Organomet. Chem.* 1990, **399**, 115–123.

1989 'Chromium Chalcogen Complexes. Insertion of Elemental Sulphur into [{Cr(cp)(CO)$_2$}$_2$Se] and [{Cr(cp)(CO)$_2$}$_2$Se$_2$] (cp = η^5-C$_5$H$_5$)', L. Y. Goh. *J. Chem. Soc., Dalton Trans.* 1989, 431–437.

'Chemistry of [{Cr(cp)(CO)$_3$}$_2$], (cp = η^5-C$_5$H$_5$). Reaction with Elemental P$_4$. A High Yield Synthesis and Crystal Structures of [{Cr(cp)(CO)$_2$}$_2$(μ-η^2-P$_2$)] and [Cr(cp)(CO)$_2$(η^3-P$_3$)]', L. Y. Goh, C. K. Chu, R. C. S. Wong, T. W. Hambley. *J. Chem. Soc., Dalton Trans.* 1989, 1951–1956.

1988 'Chemistry of [CpCr(CO)$_3$]$_2$. An Insertion Mechanism for the Formation of Cp$_2$Cr$_2$(CO)$_5$Se$_2$ and Cp$_2$Cr$_2$(CO)$_4$Se$_2$ from Cp$_2$Cr$_2$(CO)$_4$Se. Carbonylation and Crystal Structure of Cp$_2$Cr$_2$(CO)$_4$Se$_2$', W. Chen, L. Y. Goh, E. Sinn. *Organometallics* 1988, **7**, 2020–2026.

1987 'The Reaction of (Cy$_3$P)$_2$Ni(H)(CH$_3$) with Carbon Dioxide. Formation of a Hydridonickel Formate Complex, HNi(O$_2$CH)(Cy$_3$P)$_2$', D. J. Darensbourg, M. Y. Darensbourg, L. Y. Goh, M. Ludvig, P. Wiegreffe. *J. Am. Chem. Soc.* 1987, **107**, 7539–7540.

'Chemistry of [CpCr(CO)$_3$]$_2$. Synthesis of Cp$_2$Cr$_2$(CO)$_4$S, Cp$_2$Cr$_2$(CO)$_4$S$_2$, and Cp$_2$Cr$_2$(CO)$_5$S$_2$. Crystal Structure and Reactivity of Cp$_2$Cr$_2$(CO)$_4$S$_2$ and Cp$_2$Cr$_2$(CO)$_5$S$_2$', L. Y. Goh, T. W. Hambley, G. B. Robertson. *Organometallics* 1987, **6**, 1051–1057.

1986 'Photoinduced Insertion of Sulphur Bridges into a Cr≡S≡Cr Multiple Bond, and X-Ray Crystal Structure of (η^5-C$_5$H$_5$)$_2$Cr$_2$(μ-S)$_2$(μ-S$_2$)', L. Y. Goh, T. C. W. Mak. *J. Chem. Soc., Chem. Commun.* 1986, 1474–1475.

'σ-Bonded Organochromium(III) Complexes. Part 3. Decomposition in Acid Solution of Chromium(III)

Complexes containing Pyridylmethyl and Polydentate Amine Ligands', K. Crouse, L. Y. Goh. *Inorg. Chem.* 1986, **25**, 478–484.

1985 'Mono- and Di-selenium Complexes of Chromium. Syntheses and Crystal Structures of $(\eta^5\text{-}C_5H_5)_2Cr_2(CO)_4Se$ and $(\eta^5\text{-}C_5H_5)_2Cr_2(CO)_4Se_2$', L. Y. Goh, C. Wei, E. Sinn. *J. Chem. Soc., Chem. Commun.* 1985, 462–464.

1983 'Sulphur Chromium Complexes: Syntheses and Crystal Structures of $(\eta^5\text{-}C_5H_5)_2Cr_2(CO)_4S$ and $(\eta^5\text{-}C_5H_5)_2Cr_2(CO)_5S_2$', L. Y. Goh, T. W. Hambley, G. B. Robertson. *J. Chem. Soc., Chem. Commun.* 1983, 1458–1460.

1982 'Preparation and Isolation of Amine(organo)-chromium(III) Complexes in Alcoholic Media', K. Crouse, L. Y. Goh. *Inorg. Chim. Acta* 1982, **60**, 205–212.

1980 'Reactivity of Bridgehead Halides with Pentacyanocobaltate(II)', S. H. Goh, L. Y. Goh. *J. Chem. Soc., Dalton Trans.* 1980, 1641–1645.

1972 'Formation and Kinetics of Decomposition of Monoaquobis (ethylenediamine)(pyridylmethyl)chromium(III) Complexes', C. T. Loo, L. Y. Goh, S. H. Goh. *J. Chem. Soc., Dalton Trans.* 1972, 585–589

'1-Adamantylpentacyanocobaltate(III)', S. H. Goh, L. Y. Goh. *J. Organomet. Chem.* 1972, **43**, 401–403.

'Oxidation of Cobalt(I) Carbonyl Complexes and Cobalt(I)-Catalyzed Oxidation of Carbon Monoxide', J. E. Bercaw, L. Y. Goh, J. Halpern. *J. Am. Chem. Soc.* 1972, **94**, 6534–6536.

1970 'α-Halobenzylzinc Halides', S. H. Goh, L. Y. Goh. *J. Organomet. Chem.* 1970, **23**, 5–8.

'Phenylation of Azobenzene', J. Miller, D. B. Paul, L. Y. Wong, A. D. Kelso (in part). *J. Chem. Soc. (B)* 1970, 62–65.

1969 'The Reactions of Pentacyanocobaltate(II) with Hydrogen Peroxide, Hydroxylamine, and Cyanogen Iodide', P. B. Chock, R. B. K. Dewar, J. Halpern, L. Y. Wong. *J. Am. Chem. Soc.* 1969, **91**, 82–84.

1968 'Kinetics of the Addition of Hydridopentacyanocobaltate(III) to Some α,β-Unsaturated Compounds', J. Halpern, L. Y. Wong, *J. Am. Chem. Soc.* 1968, **90**, 6665–6669.

'σ-Bonded Organotransition-metal Ions. Part VI. Kinetics and Mechanism of Insertion of Hydrogen Isocyanide in

Selected List of Journal Publications

Organopentacyanocobaltate(III) Ions', M. D. Johnson, M. L. Tobe, L. Y. Wong. *J. Chem. Soc. (A)* 1968, 923–928.

'σ-Bonded Organotransition-metal Ions. Part VII. Kinetics and Mechanism of the Acid- and Base-catalysed Decomposition of the Secondary α-(2-Pyridio)-ethylpentacyanocobaltate(III) Ion', M. D. Johnson, M. L. Tobe, L. Y. Wong. *J. Chem. Soc. (A)* 1968, 929–933.

1967 'σ-Bonded Organotransition-metal Ions. Part IV. Kinetics and Mechanism of the Acid-catalysed Decomposition of the 4-Pyridiomethylpentacyanocobaltate(III) Ion', M. D. Johnson, M. L. Tobe, L. Y. Wong. *J. Chem. Soc. (A)* 1967, 491–497.

'Insertion of Hydrogen Isocyanide in Acid-catalysed Reactions of Organopentacyanocobaltate(III) Ions', M. D. Johnson, M. L. Tobe, L. Y. Wong. *J. Chem. Soc., Chem. Commun.* 1967, 298–299.

1965 'Preparation of a Complex Containing a Secondary Carbon-Chromium Bond and of Air-Stable Complexes Containing Primary Carbon-Chromium Bonds', R. G. Coombes, M. D. Johnson, M. L. Tobe, N. Winterton, L. Y. Wong. *J. Chem. Soc., Chem. Comm.* 1965, 251.

'The Addition of Benzyl Radicals to Olefinic Systems', R. L. Huang, H. H. Lee, L. Y. Wong. *J. Chem. Soc.* 1965, 6730–6737.

APPENDIX III

Published Conference Abstracts and Proceedings

1999 '*Ruthenium Compounds for 'Electric Windows'*, Proceedings of the Malaysian Chemical Congress '97, 18–20 Nov. 1997, Johor Bahru, Malaysia, ed. Z. Zakaria and Laily b. Din, on 'Chemistry and its Socio-Economic Responsibility', (April 1999), p. 385–388.

'*Breaking Invisible Barriers*', in a report of the Workshop on 'Development and Enhancement of Research and Scientific Culture', organised by Akademi Sains Malaysia and the Science and Technology Commission of the National Council of Women's Organisation (NCWO) Malaysia, November 4, 1999.

1998 '*Smart Windows — Major Energy Savings for the Built Environment in the Tropics*', L. Y. Goh, G. E. Tulloch. *Proceedings, Conference on Buildings and the Environment in Asia*, 11–13 Feb. 1998, School of Building and Real Estate, National University of Singapore, Singapore (p. 88–93).

'*Materials & Processing for Electrochromic and Solar-Power Windows*', L. Y. Goh, L. Spiccia, G. E. Tulloch. *Proceedings, 4th National Symposium on Progress in Materials Research*, 27 March 1998, Institute of Materials Research & Engineering, National University of Singapore, Singapore (p.356–359).

1994 'Role of Women in Science and Technology in Universities', Proceedings of Congress on Women in Science and Technology, organised by Institute of Policy Research and National Council of Women in Malaysia, 1994, Kuala Lumpur.

1991 'Perspectives in Organochromium-Pnicogen Chemistry', L. Y. Goh. Proceedings, National IRPA Seminar, Fifth Malaysia Plan, 16–19 December, 1991, Pulau Pinang.

1989 'Triple-Decker Sandwich η^5-P_5 and η^6-P_6 Complexes of Group 6 Metals', L. Y. Goh, R. C. S. Wong. Proceedings, Malaysian Chemical Conference 1989, 24–26 October, 1989, Kuala Lumpur (p.41–45).

'The Emerging Role of Unsupported Main Group Elements as Ligands in Transition Metal Complexes, with Special Reference to the Pnicogens', L. Y. Goh. Proceedings, Second National Symposium on Organometallic and Inorganic Chemistry, 24–25 Aug, 1989, Pulau Pinang, (p.1–10).

'Cyclopentadienylchromium Complexes of Phosphorus', 27th International Coordination Chemistry Conference, 2–7 July, 1989, Gold Coast, Queensland, Australia.

1988 'Reactions of Amine(organo)chromium Complexes', K. A. Crouse, L. Y. Goh. Proceedings, First National Symposium on Organometallic and Inorganic Chemistry, 23 March 1988, Kuala Lumpur (p. 18–27).

'Cyclopentadienyl Complexes of Chromium with Chalcogens', L. Y. Goh. Proceedings, First National Symposium on Organometallic and Inorganic Chemistry, 23 March 1988, Kuala Lumpur (p. 14–17).

'Transition Metal Clusters', 1987/1988, University of Malaya Chemistry Society Annual.

1987 'Reaction of Carbon-dioxide with Hydridomethylbistricyclohexyl-phosphinenickel(0)', D. J. Darensbourg, M. Y. Darensbourg, M. Ludvig, L. Y. Goh, et al. [See Abstracts (Inor 350)] Papers of the American Chemical Society, 194, 350.

1984 'Sulphur Complexes of the Transition Metals with Special Reference to Chromium', L. Y. Goh. Proceedings, Priochem Asia 1984, Asian Chemical Conference 1984, 29–31 March, 1984, Kuala Lumpur (p.545–554).

'Reactivity Aspects of $(\eta^5\text{-}C_5H_5)_2Cr_2(CO)_6$', L. Y. Goh. Kimia, 1984, **7**, p. 13–20

'Some Aspects of the Reactivity of $(\eta^5\text{-}C_5H_5)_2Cr_2(CO)_6$', L. Y. Goh. [See Abstracts (Inor 52)] 188th American Chemical Society National Meeting, 26–31 August, 1984, Philadelphia, U.S.A.

CPSIA information can be obtained
at www.ICGtesting.com
Printed in the USA
LVHW061452180919
631476LV00002B/12/P